# INTRODUCTORY MONOGRAPHS IN MATHEMATICS

General Editor
The late A. J. Moakes M.A.

Titles in the Series:

| | |
|---|---|
| A Boolean Algebra *Abstract and Concrete* | A. B. Bowran |
| Matrices and their Applications | J. R. Branfield and A. W. Bell |
| Programming by Case Studies | O. B. Chedzoy and Sandra E. Ford |
| Mathematics for Circuits | W. Chellingsworth |
| Logic | Nick Earle |
| Studies in Structure | J. M. Holland |
| The Core of Mathematics | A. J. Moakes |
| Numerical Mathematics *Exercises in computing with a desk calculator* | A. J. Moakes |

# LOGIC

NICK EARLE
*Headmaster, Bromsgrove School*

MACMILLAN

First published 1973

*Published by*
THE MACMILLAN PRESS LTD
*London and Basingstoke*
*Associated companies in New York Toronto*
*Dublin Melbourne Johannesburg and Madras*

SBN 333 11344 6

Text set in 10/11 pt IBM Press Roman, printed by photolithography, and bound
in Great Britain at The Pitman Press, Bath.

To C. W. L. and R. W. P. – for much forbearance

# PREFACE

No originality is claimed for this work. It aims merely to sketch out, in the briefest possible compass but with the greatest rigour consistent with that compass, the logical foundations of mathematics as far as the elementary properties of the natural numbers. Most text books of mathematics take these properties as their starting point.[1] It is the purpose of mathematical logic to show that a more secure foundation is possible.

I have not attempted complete rigour. I have, for example, touched only lightly on the question of logical types and the discerning reader will recognise that this question is in fact of great importance. Indeed I have not ventured far into the field of metalogic at all, believing that the significance of the subject is inappropriate for the student for whom the book is primarily intended, that is, someone to whom logic is a wholly new discipline. Although this deficiency may give the book a rather more 'classical' than 'modern' emphasis I contend that the concerns of modern logic are the better appreciated if the classical approach is first understood. To appreciate Einstein it is helpful to have a firm grasp of Newton. I hope that, despite these shortcomings, the argument will be found satisfying so far as it goes.

It has long been my belief that the importance of Sixth Form education – and perhaps of all formal education – lies not so much in what is learnt as in the opportunities afforded to the student to explore new topics with a view to discovering his own real interests. It is now my hope that some students will be encouraged by this monograph to find and develop interest in the further pursuit of what is perhaps the most fascinating if the most elusive of all disciplines mathematical logic.

In conclusion, I should like to thank the following Institutions for permission to reproduce in the book questions taken from past examination papers:

University of Cambridge
University of Edinburgh
University of London, Schools Examinations Department
The British Computer Society.

Similar acknowledgement is also due to the publishers of:

Langer: Introduction to Symbolic Logic (Van Nostrand Reinhold Company) and

Suppes: Introduction to Logic (Dover Publications Inc., New York, 1953, 1967; reprinted through permission of the publishers)

# CONTENTS

1

# THE LOGIC OF PROPOSITIONS

## 1.1 Demonstration and proof: the law of gravity

The following extract is taken from Galileo's Dialogue concerning Two New Sciences. It deals with the question of whether or not bodies of unequal mass fall with unequal speeds. Galileo's own demonstration that in fact they do not is often taken as the starting point for a history of modern science. It may appropriately serve to introduce a discussion of the part which logic has had to play in science. The dialogue takes place between three people – Simplicio, Sagredo and Salvator.

SIMPLICIO: Aristotle's language would seem to indicate that he had tried the experiment (of allowing unequal masses to fall through equal distances) because he says 'We *see* the heavier (fall faster than the lighter)'; now the word *see* shows that he had made the experiment.

SAGREDO: But I, Simplicio, who have made the test, can assure you that a cannon-ball weighing one or two hundred pounds or even more will not reach the ground by as much as a span ahead of a musket-ball weighing only half a pound, provided both are dropped from a height of one hundred cubits.

SALVATOR: But even without further experiment is is possible to prove clearly by means of a short and conclusive argument that a heavier body does not move more rapidly than a lighter one, provided both bodies are of the same material and in short such as those mentioned by Aristotle. But tell me, Simplicio, whether you admit that each falling body acquires a definite speed fixed by nature, a velocity which cannot be increased or diminished except by the use of force or resistance.

SIMPLICIO: There can be no doubt that one and the same body, moving in a single medium, has a fixed velocity which is determined by nature and which cannot be increased except by the addition of momentum or diminished except by some resistance which retards it.

SALVATOR: If then we take two bodies whose natural speeds are different, it is clear that on uniting the two, the most rapid one will be partly retarded by the slower and the slower will be somewhat hastened by the swifter. Do you not agree with this opinion?

SIMPLICIO: You are unquestionably right.

SALVATOR: But if this is true and if a large stone moves with a speed of, say, eight (units) while a smaller moves with a speed of four then when they are united the system will move with a speed less than eight; but the two stones together make a stone larger than that which before moved with a speed of eight. Hence the heavier body moves with less speed than the lighter; an effect which is contrary to your supposition.

In this passage two approaches to truth are contrasted – the experimental and the logical. Which is to be preferred? In favour of the experimental method it is often urged that anyone, however slight his intellectual capacity, can make the experiment. Anyone who cares to do so can see for himself the truth of the assertion Galileo makes. Against the experimental method it can be argued (and from the logician's point of view the argument is inescapable) that however often the experiment is conducted there can be no certainty that on the next occasion the result will be the same. In order to be convinced even by a large number of demonstrations we have to make some assumptions about the regularity of the world; and these are not easy to justify.

In favour of the logical method it is said that once its assumptions are granted the argument moves irresistibly to its conclusion. And there are many cases – of which the foregoing might well be one since Galileo is compelled to admit that the speeds of different bodies are not *exactly* equal – when the assumptions are easier to verify than the conclusion which it is desired to prove.

In most cases, in fact, we need both experiment *and* reasoning in order to reach a conclusion. If, as I hope to show, mathematics is just a special kind of reasoning then this is one kind of reasoning which in most branches of science we cannot do without. But can we do without experiment, as Salvator seems to suggest? It was Aristotle's mistake – and that of the ancients generally[2] – to suppose that we can; that sufficient assumptions can be found which are simply *self-evident.* When these assumptions – as, for example, that the rate at which a body falls through a medium is proportional inversely to the density of the medium –

are shown to be false, we are not surprised to find the experiment demonstrates the falsehood of the conclusion also.

Equally, if experiment falsifies the conclusion of a valid argument, one or more of the assumptions on which it is based must be rejected. The purpose of logic is thus two-fold: by constructing arguments which lead to demonstrable conclusions to seek for a minimum of assumptions upon which the conclusions may be shown logically to depend; and by obtaining further conclusions from these assumptions to suggest further experiments which may, in their turn, show whether the assumptions are still tenable or not.

How the conclusions are reached from the assumptions, or **premisses**, it is now time to enquire.

## 1.2 Propositions

The first question we must ask is: what can be said of the assumptions themselves? They must be statements of fact. We cannot derive other statements of fact from expressions of opinion, from exclamations, ejaculations, imperatives – nor even from statements (so called) as to their authors' own state of mind. 'Great Scott!', 'I hope it stays fine', 'Keep quiet!', 'I am bored' are not statements in the true sense and can form no part of an argument.

A statement, or proposition[3] as we shall now term it, is characterised in three ways:

(i) it is capable of both assertion and denial;
(ii) either its assertion or denial is, at least theoretically, capable of being shown to be true, but not both;
(iii) it is capable of being conjoined with another proposition, usually by means of the conjunction 'and', to form a third.

These three characteristics are to be taken as a sufficient indication, at least for the moment, of the means whereby a proposition may be recognised.

### (a) *Truth-values*

Since every proposition is capable of being shown to be either true or false it is said to have a **truth value**. This truth-value is 'true' if the proposition is true, 'false' if it is false.[4] For convenience we denote these values by the symbols 1 and 0; in many books the symbols $T$ and $F$ are used, and any pair of distinguishable symbols would in fact suffice.

### (b) *Negation*

Since every proposition can be denied we use a common symbol, the prime, adjoined to whatever symbol denotes a proposition to denote that proposition whose assertion is equivalent to the denial of the first. Thus:

*Queen Anne is dead'*

is used to denote:

*Queen Anne is not dead*

Other symbols in common use are:

*'Queen Anne is dead'.*
*'Queen Anne is dead'*, and
~*'Queen Anne is dead'*

the quotation marks being inserted to emphasise that it is the proposition which is asserted or denied and not the sentence which happens to express that proposition in English. Each of these is termed the **negation** of *Queen Anne is dead.*

(c) *Conjunction and alternation*

Finally since any proposition can be conjoined with any other to form a third we use the symbol ∧ placed between the symbols denoting the first and second to denote the third. Thus:

*Queen Anne is dead* ∧ *William won the Battle of Hastings*

is used to denote the single proposition:

*Queen Anne is dead and William won the Battle of Hastings*

which is the **conjunction** of the other two.

Propositions can be joined in other ways. We can, for instance form the proposition whose assertion is:

*Queen Anne is dead or William won the Battle of Hastings*

The truth-value of this proposition would be unaltered even if it could be shown that William was not in fact present at the Battle of Hastings. It is a proposition, known as the **disjunction** or **alternation** of the first two which has the value 'true' if either or both of the propositions from which it is formed has the value 'true'.

This way of joining propositions for which the symbol ∨ is used is not unrelated to conjunction. The connection between them is symbolised thus:

$$(p \vee q) \equiv (p' \wedge q')' \tag{1.1}$$

The expression on the right-hand side of the sign of equivalence, ≡, denotes the negation of $p' \wedge q'$, the brackets indicating that the whole of $p' \wedge q'$ is being negated. The sign of equivalence serves to denote the fact that the propositions connected by it have the same truth-value, *whatever the truth values of p and q may be*, $p$ and $q$ denoting any propositions at all.

The truth of the whole proposition[5] expressed by (1.1) above can now be verified. All we need to do is to consider every possible combination of truth and falsehood in $p$ and $q$ (there are only four of them) and enquire whether for every such combination the expressions on each side of the sign of equivalence have the same truth-value.

(d) *Truth-tables*

The method employed is that of the **truth-table**, a table which sets out in a regular way, the truth values of a complex proposition or **formula**, for all possible combinations of the truth values of its parts. The tables relevant to (1.1) are given in Table 1.1.

Table 1.1. $p \wedge q$

| $p$ | $q$ | $p \wedge q$ |
|---|---|---|
| 1 | 1 | 1 |
| 1 | 0 | 0 |
| 0 | 1 | 0 |
| 0 | 0 | 0 |

Table 1.1 tells us that $p \wedge q$ has the value 'true' if $p$ and $q$ both have the value 'true' but has the value 'false' otherwise.

Table 1.2. $p \vee q$

| $p$ | $q$ | $p \vee q$ |
|---|---|---|
| 1 | 1 | 1 |
| 1 | 0 | 1 |
| 0 | 1 | 1 |
| 0 | 0 | 0 |

Table 1.2 tells us that $p \vee q$ has the value 'false' if $p$ and $q$ both have the value 'false', but has the value 'true' otherwise.

Table 1.3. $(p' \wedge q')'$

| $p$ | $q$ | $p'$ | $q'$ | $(p' \wedge q')$ | $(p' \wedge q')'$ |
|---|---|---|---|---|---|
| 1 | 1 | 0 | 0 | 0 | 1 |
| 1 | 0 | 0 | 1 | 0 | 1 |
| 0 | 1 | 1 | 0 | 0 | 1 |
| 0 | 0 | 1 | 1 | 1 | 0 |

Finally, Table 1.3 (in which the third and fourth columns are obtained by changing the truth-values in the first and second; and the sixth by changing those in the fifth), tells us that $(p' \wedge q')'$ has the same truth-value as $p \vee q$ for each corresponding combination of the truth-values of $p$ and $q$ and is therefore its equivalent.

This means that wherever $(p' \wedge q')'$ occurs we may substitute $p \vee q$ for it, a rule which allows us, if we wish, to take the former as a *definition* of the latter. For the moment, however, we limit ourselves to defining $p \vee q$ simply by means of Table 1.2.

(e) *Nand and nor*

We have just shown that a separate truth-table for alternation is unnecessary if the tables for negation and conjunction are given. But we could go further and obtain all the possible combinations of propositions from a single table.

Table 1.4. $p \uparrow q$

| $p$ | $q$ | $p \uparrow q$ |
|---|---|---|
| 1 | 1 | 0 |
| 1 | 0 | 0 |
| 0 | 1 | 0 |
| 0 | 0 | 1 |

Consider Table 1.4, which gives the truth-value of the proposition '*p*-nor-*q*', which is 'true' only if neither that of $p$, nor that of $q$ is true. The symbol ↑ is one of the so-called **Sheffer stroke-functions** after Sheffer, an American logician who published the table in 1913.

The other stroke-function, ↓, is used in the formation of '*p*-nand-*q*', a proposition which has the value 'false' only when both $p$ and $q$ have the value 'true'.[6] (Table 1.5.)

Table 1.5. $p \downarrow q$

| $p$ | $q$ | $p \downarrow q$ |
|---|---|---|
| 1 | 1 | 0 |
| 1 | 0 | 1 |
| 0 | 1 | 1 |
| 0 | 0 | 1 |

It is not difficult to verify that all the other tables – for negation, conjunction and alternation, can be derived from either one of these tables and in Exercise 1.1 you are invited to do this for yourself.

**Exercise**

1.1 Use truth-tables to verify the following:

(a) $p' \equiv p \uparrow p$ (b) $p' \equiv p \downarrow p$
(c) $(p \vee q) \equiv ((p \uparrow q) \uparrow (p \uparrow q))$ (d) $(p \wedge q) \equiv ((p \downarrow q) \downarrow (p \downarrow q))$
(e) $(p \wedge q) \equiv ((p \uparrow p) \uparrow (q \uparrow q))$ (f) $(p \vee q) \equiv ((p \downarrow p) \downarrow (q \downarrow q))$

It is noticeable that these equivalences exhibit a special kind of symmetry. We can interchange the stroke functions without affecting the rule for negation. But if we interchange them elsewhere we must also interchange the symbols for conjunction and alternation. This is referred to as the principle of **duality**, and (b), (d) and (f) are the duals of (a), (c) and (e).

## 1.3 Formulae

(a) *Implication*

Another truth-table with which we shall be a good deal concerned is Table 1.6, which demonstrates the important fact that when $p$, $p' \vee q$

Table 1.6. $p' \vee q$

| $p$ | $q$ | $p'$ | $p' \vee q$ |
|---|---|---|---|
| 1 | 1 | 0 | 1 |
| 1 | 0 | 0 | 0 |
| 0 | 1 | 1 | 1 |
| 0 | 0 | 1 | 1 |

have the value 'true', $q$ has the value 'true' also. For this reason $p' \vee q$ is often referred to as 'if-$p$-then-$q$' and is denoted by $p \Rightarrow q$ or, less frequently, by $q \Leftarrow p$. If it is known to have the value 'true' we are able to 'infer' the assertion of $q$ from the assertion of $p$ (see section **1.4**(a).

This answers to the ordinary usage of the word 'inference' which allows us to infer 'the day is warm' from 'the sun is shining' if we already know that either the sun is not shining or the day is warm (or both).

You may have noticed that a few lines back it was asserted that when $p, p' \vee q$ have the value 'true', $q$ has the value 'true'. This might well have been expressed as 'if $p, p' \vee q$ have the value 'true', then $q$ has the value 'true'. This suggests another 'if-then . . .' proposition, namely:

'if if-$p$-then-$q$ and $p$ then $q$'

or, in symbols:

$$((p \Rightarrow q) \wedge p) \Rightarrow q \qquad (1.2)$$

and this proposition, unlike $p \Rightarrow q$ itself, is 'true' whatever the truth-values of $p$ and $q$ may be, as can be seen from Table 1.7.

Table 1.7.

| $p$ | $q$ | $p \Rightarrow q$ | $(p \Rightarrow q) \wedge p$ | $((p \Rightarrow q) \wedge p)'$ | $((p \Rightarrow q) \wedge p)' \vee q$ |
|---|---|---|---|---|---|
| 1 | 1 | 1 | 1 | 0 | 1 |
| 1 | 0 | 0 | 0 | 1 | 1 |
| 0 | 1 | 1 | 0 | 1 | 1 |
| 0 | 0 | 1 | 0 | 1 | 1 |

The formal title of the connection denoted by $\Rightarrow$ is 'material implication';[7] as the use of the arrow suggests it is a directed connection unlike $\uparrow, \downarrow, \wedge$, and $\vee$, so that $p \Rightarrow q$ has not always the same truth-value as $q \Rightarrow p$. And it is often easier to refer to it by the word 'implies' than by the words 'if-then'.

(b) *Equivalence*

Another important way of connecting propositions is by **equivalence**. I have already said that two propositions are equivalent when their truth-values are the same, i.e., they are 'true' or 'false' together. Table 1.8 shows that this is the same as saying that each implies the other. The equivalence of $p$ and $q$, symbolised by $p \equiv q$ (or sometimes by $p \Leftrightarrow q$ if mutual implication is to be stressed) is also referred to as '$p$ if-and-only-if $q$' or '$p$ iff $q$'.

Table 1.8. $p \equiv q$

| $p$ | $q$ | $p \Rightarrow q$ | $q \Rightarrow p$ | $(p \Rightarrow q) \wedge (q \Rightarrow p)$ |
|---|---|---|---|---|
| 1 | 1 | 1 | 1 | 1 |
| 1 | 0 | 0 | 1 | 0 |
| 0 | 1 | 1 | 0 | 0 |
| 0 | 0 | 1 | 1 | 1 |

(c) *Tautology*

Expression (1.2) is a proposition whose value is 'true' whatever the values of its constituent propositions, $p$ and $q$. Such a proposition is called a **tautology**.

Other examples are:

$$p \vee p' \qquad p \equiv p \qquad p \equiv (p')' \qquad p \equiv p \vee p \qquad p \equiv p \wedge p \quad (1.3)$$

When expressed in words these propositions have a tautologous sound. But sound is not a good guide. Consider the following example:

$$(p \vee (q \equiv r)) \equiv ((p \vee q) \equiv (p \vee r)) \qquad (1.4)$$

This does not sound obviously tautological. The only way of making sure that it is so is by constructing a truth-table (see Table 1.9).

Table 1.9.

| $p$ | $q$ | $r$ | $p \vee q$ | $p \vee r$ | $(p \vee q) \equiv (p \vee r)$ | $p \vee (q \equiv r)$ |
|---|---|---|---|---|---|---|
| 1 | 1 | 1 | 1 | 1 | 1 | 1 |
| 1 | 1 | 0 | 1 | 1 | 1 | 1 |
| 1 | 0 | 1 | 1 | 1 | 1 | 1 |
| 1 | 0 | 0 | 1 | 1 | 1 | 1 |
| 0 | 1 | 1 | 1 | 1 | 1 | 1 |
| 0 | 1 | 0 | 1 | 0 | 0 | 0 |
| 0 | 0 | 1 | 0 | 1 | 0 | 0 |
| 0 | 0 | 0 | 0 | 0 | 1 | 1 |

We could now construct three further columns, one for $(p \vee (q \equiv r)) \Rightarrow ((p \vee q) \equiv (p \vee r))$, one for the reverse implication and one for the conjunction of the two and obtain nothing but 1s in the final column. In fact, the correspondence of 1s and 0s in the last two columns of

Table 1.9 together with Table 1.7, is sufficient to demonstrate that (1.4) is indeed a tautology.

(d) *List of tautologies*

For purposes of reference a list of useful tautologies is appended here.

Expressions (1.8)–(1.10) are the **commutative rules** for conjunction, alternation and equivalence. They allow us to interchange propositions appearing on either side of the symbols $\wedge, \vee$ and $\equiv$ without affecting truth-values:

$$(p')' \equiv p \tag{1.5}$$
$$(p \wedge q) \Rightarrow p \tag{1.6}$$
$$p \Rightarrow p \vee q \tag{1.7}$$
$$(p \wedge q) \equiv (q \wedge p) \tag{1.8}$$
$$(p \vee q) \equiv (q \vee p) \tag{1.9}$$
$$(p \equiv q) \equiv (q \equiv p) \tag{1.10}$$

Expressions (1.11) and (1.12) are De Morgan's **Laws**. The first of these is (1.1) with $p'$ for $p$ and $q'$ for $q$. The second is the dual of the first:

$$(p \wedge q)' \equiv (p' \vee q') \tag{1.11}$$
$$(p \vee q)' \equiv (p' \wedge q') \tag{1.12}$$
$$(p \equiv q)' \equiv (p \equiv q') \tag{1.13}$$
$$(p \Rightarrow q)' \equiv (p \wedge q') \tag{1.14}$$
$$(p \Rightarrow q) \equiv (q' \Rightarrow p') \tag{1.15}$$
$$(p \equiv q) \equiv (p' \equiv q') \tag{1.16}$$

Expressions (1.17) and (1.18) are the **associative rules** for conjunction and alternation. They allow us to write without ambiguity $(p \wedge q \wedge r)$ in place of the expression on either side of the first equivalence, and $(p \vee q \vee r)$ in place of the expression on either side of the second. It is not difficult to see that these principles of multiple conjunction and multiple alternation can be extended to any number of propositions:

$$(p \wedge (q \wedge r)) \equiv ((p \wedge q) \wedge r) \tag{1.17}$$
$$(p \vee (q \vee r)) \equiv ((p \vee q) \vee r) \tag{1.18}$$

Expressions (1.19) – (1.22) are the **distributive rules**, of alternation over conjunction, of conjunction over alternation, and of each over itself. They are not the only distributive rules, however. Alternation, for example, is distributive over equivalence (1.4).

$$(p \vee (q \wedge r)) \equiv ((p \vee q) \wedge (p \vee r)) \tag{1.19}$$
$$(p \wedge (q \vee r)) \equiv ((p \wedge q) \vee (p \wedge r)) \tag{1.20}$$
$$(p \wedge (q \wedge r)) \equiv ((p \wedge q) \wedge (p \wedge r)) \tag{1.21}$$
$$(p \vee (q \vee r)) \equiv ((p \vee q) \vee (p \vee r)) \tag{1.22}$$

Expressions (1.23) and (1.24) are the **rules of absorption.**

$$(p \wedge (p \vee q)) \equiv p \qquad (1.23)$$
$$(p \vee (p \wedge q)) \equiv p \qquad (1.24)$$

The remainder are a miscellany, but all are made use of at least once in what follows.

$$(p \Rightarrow q) \Rightarrow ((p \wedge r) \Rightarrow (q \wedge r)) \qquad (1.25)$$
$$(p \Rightarrow q) \Rightarrow ((p \vee r) \Rightarrow (q \vee r)) \qquad (1.26)$$
$$(p \Rightarrow r) \wedge (q \Rightarrow r) \Rightarrow ((p \vee q) \Rightarrow r \qquad (1.27)$$
$$((p \equiv q) \wedge (r \equiv s)) \Rightarrow ((p \wedge r) \equiv (q \wedge s)) \qquad (1.28)$$
$$((p \equiv q) \wedge (r \equiv s)) \Rightarrow ((p \vee r) \equiv (q \vee s)) \qquad (1.29)$$
$$((p \equiv q) \wedge (r \equiv s)) \Rightarrow ((p \equiv r) \equiv (q \equiv s)) \qquad (1.30)$$
$$(q \vee (p \wedge p')) \equiv q \qquad (1.31)$$
$$(q \wedge (p \wedge p')) \equiv p \wedge p' \qquad (1.32)$$
$$(q \vee (p \vee p')) \equiv p \vee p' \qquad (1.33)$$
$$(q \wedge (p \vee p')) \equiv q \qquad (1.34)$$
$$(q \wedge q') \equiv (p \wedge p') \qquad (1.35)$$
$$(q \vee q') \equiv (p \vee p') \qquad (1.36)$$
$$(p \Rightarrow q) \equiv (p \equiv (p \wedge q)) \qquad (1.37)$$
$$(p \Rightarrow q) \equiv (q \equiv (p \vee q)) \qquad (1.38)$$
$$((p \Rightarrow q) \wedge (p \Rightarrow r)) \equiv (p \Rightarrow (q \wedge r)) \qquad (1.39)$$
$$(p \Rightarrow p') \Rightarrow p' \qquad (1.40)$$
$$p' \wedge (q \Rightarrow p) \Rightarrow q' \qquad (1.41)$$
$$p \wedge (p \Rightarrow q) \Rightarrow q \qquad (1.42)$$
$$p \Rightarrow (q \Rightarrow (p \wedge q)) \qquad (1.43)$$
$$((p \wedge q) \Rightarrow r) \Rightarrow (p \Rightarrow (q \Rightarrow r)) \qquad (1.44)$$
$$((p \Rightarrow q) \wedge (p \vee r)) \Rightarrow (q \vee r) \qquad (1.45)$$
$$((p \equiv q) \wedge (q \equiv r)) \Rightarrow (p \equiv r) \qquad (1.46)$$
$$(p \Rightarrow (p \wedge q)) \equiv (p \Rightarrow q) \qquad (1.47)$$
$$(p \equiv q)' \equiv ((p \wedge q') \vee (p' \wedge q)) \qquad (1.48)$$
$$(q \equiv (p \vee p')) \equiv q \qquad (1.49)$$
$$(q \equiv (p \wedge p')) \equiv q' \qquad (1.50)$$
$$p \Rightarrow (q \vee q') \qquad (1.51)$$
$$((p \vee q) \Rightarrow r) \Rightarrow (p \Rightarrow r) \qquad (1.52)$$
$$q \Rightarrow (p \Rightarrow q) \qquad (1.53)$$
$$((p \wedge q) \Rightarrow r) \wedge ((p \wedge r) \Rightarrow q)) \equiv ((p \wedge q) \equiv (p \wedge r)) \qquad (1.54)$$
$$((p \Rightarrow q) \wedge (q \equiv p)') \Rightarrow (q \Rightarrow p)' \qquad (1.55)$$

---

**Exercise**

1.2 Use truth-tables to verify the tautologies listed in section **1.3**(c). Note that it is not always necessary to write out the truth tables in full. It is plain, for example, in (1.21) that $(p \wedge (q \wedge r))$ can only have the value 'true' when $p$, $q$ and $r$ all have this value. Similarly $p \vee (q \vee r)$ can only have the value 'false' when $p$, $q$ and $r$ all have this value. And so on.

---

## 1.4 Argument

Logic has traditionally concerned itself with the question: 'under what circumstances may an argument be said to be valid, i.e., how may a conclusion fairly be reached from given premisses, regardless of the truth or falsehood of those premisses?' This section gives a preliminary answer to that question, a slightly more general answer being deferred to section **1.6**. First, however, we examine the justification for each individual step in the argument, i.e., we define a valid inference.

(a) *Inference*

Among the tautologies, special importance is attached to those whose main connective is a material implication and in which either a single (though not necessarily a simple) proposition occurs on the left-hand side of this connective or a conjunction of propositions. Such a tautology will be of the form:

$$p \Rightarrow q$$

or

$$(p \wedge q) \Rightarrow r$$

and (1.40) – (1.45) are instances.

In such a case, the assertion[8] of the proposition on the right of the sign of implication is said to be a **valid inference** from the assertion of the proposition or propositions conjoined on the left. The latter are referred to as the **premisses** of the inference and the former as the **conclusion.**

Some instances of valid inference are given below.

(i) The rule of detachment

The relevant tautology is:

$$((p \Rightarrow q) \wedge p) \Rightarrow q$$

From the premisses $p \Rightarrow q$ and $p$ we infer $q$.

*Example 1*

Premiss: *Birds are not mammals implies birds lay eggs*
Premiss: *Birds are not mammals*
Conclusion: *Birds lay eggs*

Here $p$ stands for 'birds are not mammals' and $q$ for 'birds lay eggs'. Notice that we cannot infer either of the premisses from the conclusion.

*Example 2*

Premiss: *The platypus is not a mammal implies the platypus lays eggs*
Premiss: *The platypus is not a mammal*
Conclusion: *The platypus lays eggs*

Here $p$ stands for 'the platypus is not a mammal' and $q$ for 'the platypus lays eggs'. Notice that though the inference is valid and the conclusion true, one premiss is in fact false.

*Example 3*

Premiss: *The platypus is a mammal implies the platypus does not lay eggs*
Premiss: *The platypus is a mammal*
Conclusion: *The platypus does not lay eggs*

Here $p$ stands for 'the platypus is a mammal' and $q$ for 'the platypus does not lay eggs'. Again the inference is valid and one premiss is false, but in this case the conclusion is false also.

(ii) The syllogism

The relevant tautology is:

$$((p \Rightarrow q) \wedge (q \Rightarrow r)) \Rightarrow (p \Rightarrow r)$$

(The work of verifying that this is in fact a tautology may be shortened by observing that it can only be 'false' in the case where $p \Rightarrow r$ is 'false'; so the only two cases we need consider are those in which $p$ is 'true' and $r$ is 'false'.)

From the premisses $p \Rightarrow q$ and $q \Rightarrow r$ we infer $p \Rightarrow r$.

*Example 4*

Premiss: *He is a grandfather implies he is a father*
Premiss: *He is a father implies he is adult*
Conclusion: *He is a grandfather implies he is adult*

Here $p$ stands for 'he is a grandfather', $q$ for 'he is a father' and $r$ for 'he is adult'.

(iii) The sorites

The relevant tautology is:

$$((p \Rightarrow q) \wedge (q \Rightarrow r) \wedge (r \Rightarrow s)) \Rightarrow (p \Rightarrow s)$$

This is an extension of the syllogism, and the conclusion is more easily reached as the result of a valid argument than by a single valid inference (section **1.4**(c)), since the verification of the tautology involves the examination of eight cases.

*Example 5*

Premiss: *The sun is visible implies the sun is up*
Premiss: *The sun is up implies my watch has stopped*
Premiss: *My watch has stopped implies I forgot to wind it*
Conclusion: *The sun is visible implies I forgot to wind my watch*

Here $p$ stands for 'the sun is visible', $q$ for 'the sun is up', $r$ for 'my watch has stopped' and $s$ for 'I forgot to wind it'.

(b) *Rules of inference*

The definition of a valid inference given in section **(1.4(a))** suggest the following **rules of inference**:

| (i) | (ii) | (iii) | (iv) |
|---|---|---|---|
| $\dfrac{(p \vee p') \Rightarrow q}{q}$ | $\dfrac{\begin{matrix} p \\ q \end{matrix}}{p \wedge q}$ | $\dfrac{\begin{matrix} p \Rightarrow q \\ p \end{matrix}}{q}$ | $\dfrac{\begin{matrix} p \Rightarrow q \\ q \Rightarrow r \end{matrix}}{p \Rightarrow r}$ |

The horizontal line is here taken to mean that the assertion of the proposition or propositions above it allow us to proceed to the assertion of the proposition below it. In the ensuing pages this procedure will often be followed, though the horizontal line will generally be omitted. Its use will be justified by the bracketed letters (RI) on the right of the conclusion, together with a reference by a similarly bracketed figure to the premisses made use of. That the rule of inference is valid may be verified by forming the alternation of the conclusion with the negated conjunction of the premisses and testing for a tautology. Where the rule of inference used is one of those listed above the letters (RI) may also be omitted.

(c) *Valid argument*

We are now in a position to give a formal definition of a **valid argument**, namely: a list of assertions each of which is *either* a premiss of the argument *or* is inferred from one or more previous assertions in the list by means of a rule of inference *or* is equivalent to a previous assertion in the list by definition. The last assertion in the list is the **conclusion of the argument.**

As an example of an argument we reach the conclusion of the sorites from its premisses as follows:

| | | |
|---|---|---|
| $p \Rightarrow q$ | Premiss | (1.56) |
| $q \Rightarrow r$ | Premiss | (1.57) |
| $p \Rightarrow r$ | (RI) (1.56) and (1.57) | (1.58) |
| $r \Rightarrow s$ | Premiss | (1.59) |
| $p \Rightarrow s$ | (RI) (1.58) and (1.59) | (1.60) |

**Exercises**

Establish the following tautologies:

1.3 $(q \uparrow (p \uparrow (p \uparrow q))) \equiv (q \uparrow q)$
1.4 $(q \downarrow (p \downarrow (p \downarrow q))) \equiv (q \downarrow q)$
1.5 $(p \vee (q \Rightarrow r)) \equiv ((p \vee q) \Rightarrow (p \vee r))$
(Alternation distributes over implication.)

Establish the following tautologies and give examples of inferences justified by them:

1.6 $((p \Rightarrow q) \wedge q') \Rightarrow p'$
1.7 $(((p \wedge r') \Rightarrow q') \wedge (p \wedge q)) \Rightarrow r$
1.8 $((p \Rightarrow q) \wedge p) \Rightarrow (q' \Rightarrow r)$

What conclusions may be reached from the given premisses in the following cases?

1.9 Premiss: *If yesterday was Monday, today is Tuesday*
Premiss: *If today is Tuesday, tomorrow is Wednesday*
Premiss: *Tomorrow is not Wednesday*

1.10 Premiss: *Either prices do not fall or there is no over production*
Premiss: *Either there is over production or factories close*
Premiss: *If factories close then there is social unrest*

## 1.5 The construction of a proposition with given properties

It is sometimes useful to be able to construct a proposition which shall be true in some circumstances and false in others. A classic instance is provided by the problem of the snake in the box.[9] A prisoner is locked in a cell with two boxes, identified as 'the left-hand box' and 'the right-hand box'. He knows that in one box is the key that will let him out of the cell; in the other is a deadly snake. His captor (who knows which is which) has promised to answer one question; he does not undertake to answer truthfully, but he does undertake to answer in a manner which is not self-contradictory. What question should the prisoner ask?

The problem is to construct a proposition whose assertion – even whose false assertion – will be possible only if the key is in, say, the left-hand box, and whose denial – even whose false denial – will be possible only if it is in the right-hand box. There are four possibilities to be considered, corresponding to two propositions either of whose assertions may be true or false. The first proposition is 'the key is in the left-hand box'; the second is 'the captor (in answering the question) is telling the truth'.

Suppose $p$ to stand for the first proposition and $q$ for the second; and let $r$ stand for the proposition which the captive is to put. We can then draw up a truth-table as in Table 1.10.

Table 1.10.

| $p$ | $q$ | Answer required | $r$ |
|---|---|---|---|
| 1 | 1 | $r$ is 'true' | 1 |
| 1 | 0 | $r$ is 'true' | 0 |
| 0 | 1 | $r$ is 'false' | 0 |
| 0 | 0 | $r$ is 'false' | 1 |

In this table the third column is obtained from the first in the obvious manner since we want '$r$ is 'true' if and only if the key is indeed in the left-hand box whether or not the captor is speaking the truth. The fourth column is then constructed so that the second becomes the equivalence of the third and fourth, i.e., the captor is telling the truth when the third is equivalent to the fourth and lying when it is not. We now seek a proposition that will take the values indicated in the fourth column for the appropriate values of $p$ and $q$. Clearly $r$ is the equivalence of $p$ and $q$, i.e.:

$$r \equiv ((p \wedge q) \vee (p' \wedge q'))$$

and the question which the prisoner should put is:

*Is it true that* either *the key is in the left-hand box and you are telling the truth* or *the key is in the right-hand box and you are lying?*

*Conjunctive and disjunctive normal form*

The last step in the solution of this problem, the construction of $r$ so that it shall have the prescribed truth-values for all possible combinations of the truth-values of $p$ and $q$, was in this case simple. It is not always so. Nor is the converse procedure, that of determining for precisely which combinations of truth-values in its component propositions the 'truth' or 'falsehood' of a complex proposition is secured. Here are some examples to illustrate the general procedure in such a case.

*Example 6*

*For what combinations of the truth-values of p, q and r is the proposition* $(p \vee q \vee r) \wedge (p' \vee q')$ *'false'?*

In the first place you will notice that the second bracket lacks any mention of $r$. Such a mention can be introduced without affecting the truth-value of the proposition since, $r \wedge r'$ being always 'false', the contents of this bracket are equivalent to $p' \vee q' \vee (r \wedge r')$ which in turn is equivalent to $(p' \vee q' \vee r) \wedge (p' \vee q' \vee r')$ by the distributive law for alternation over conjunction (1.19).

So we infer successively:

$$(p \vee q \vee r) \wedge (p' \vee q') \equiv (p \vee q \vee r) \wedge (p' \vee q' \vee (r \wedge r')) \qquad \text{(1.28) and (1.31)} \qquad (1.61)$$

$$(p \vee q \vee r) \wedge (p' \vee q' \vee (r \wedge r')) \equiv (p \vee q \vee r) \wedge (p' \vee q' \vee r) \wedge (p' \vee q' \vee r') \qquad \text{(1.28) and (1.19)} \qquad (1.62)$$

$$(p \vee q \vee r) \wedge (p' \vee q') \equiv (p \vee q \vee r) \wedge (p' \vee q' \vee r) \wedge (p' \vee q' \vee r') \qquad \text{(1.61), (1.62) and (1.46)} \qquad (1.63)$$

Expression (1.63) asserts the equivalence of the original proposition to another, namely $(p \vee q \vee r) \wedge (p' \vee q' \vee r) \wedge (p' \vee q' \vee r')$, which is said to be in **full conjunctive normal form** in $p, q, r$. This means that it is a multiple conjunction of multiple disjunctions, each containing either $p, q, r$ or its negation.

The advantage of expressing a proposition in this form becomes clear when we notice that for the proposition itself to be 'false' it is necessary for at least one of the disjunctions to be 'false'; and that for any disjunction to be 'false' it is necessary for all its constituent propositions to be 'false'.

Thus the original proposition $(p \vee q \vee r) \wedge (p' \vee q')$ is 'false' if:

either $p, q, r$ are all 'false' (first disjunction)
or $p', q', r$ are all 'false' (second disjunction)
or $p', q', r'$ are all 'false' (third disjunction)

However we have not yet demonstrated that this exhausts all the possibilities. That such is the case emerges from the following considerations.

In a full conjunctive normal form, no two of the propositions conjoined can be 'false' for the same set of truth-values in the constituent propositions, since for any disjunction to be 'false' all its constituents must be 'false'; and in any other disjunction at least one of these has been negated. Thus no combination of truth-values other than those represented by the 'falsehood' of the several disjunctions in a full conjunctive normal form can make the whole proposition 'false'.

Since an extension of the method employed in this example shows that every proposition can be expressed in full conjunctive normal form we may state the following principle:

*Every proposition involving a given number of constituent propositions can be expressed in full conjunctive normal form in one and only one way.*

*Example 7*

*For what combination of the truth-values of p, q and r is the proposition* $(p \vee q \vee r) \wedge (p' \vee q')$ *'true'?*

In the light of the principle just stated we can answer straight away 'all combinations except $p, q, r$ all 'false', $p', q', r$ all 'false'; and $p', q', r'$ all 'false' '. However in order to illustrate the method of the full disjunctive normal form we proceed in a slightly different way.

By making repeated use of (1.19) and (1.20) we reach the following:

$$((p \vee q \vee r) \wedge (p' \vee q'))$$
$$\equiv ((p \wedge p') \vee (q \wedge p') \vee (r \wedge p') \vee (p \wedge q') \vee (q \vee q') \vee (r \wedge q'))$$

Now $(p \wedge p')$, $(q \wedge q')$ are always 'false' and can therefore be omitted from any multiple disjunction (1.31). Thus we reach:

$$((p \vee q \vee r) \wedge (p' \vee q')) \equiv ((q \wedge p') \vee (r \wedge p') \vee (p \wedge q') \vee (r \wedge q'))$$

As before, conjunctions which lack a mention of any of the constituent propositions can have this deficiency remedied by conjunction with an appropriate tautology so that, for example, $(q \wedge p')$ can be replaced by $(q \wedge p' \wedge (r \vee r'))$ or by its equivalent $((q \wedge p' \wedge r) \vee (q \wedge p' \wedge r'))$.

In this way the original proposition can be shown to be equivalent to:

$$(q \wedge p' \wedge r) \vee (q \wedge p' \wedge r') \vee (r \wedge p' \wedge q) \vee (r \wedge p' \wedge q')$$
$$\vee (p \wedge q' \wedge r) \vee (p \wedge q' \wedge r') \vee (r \wedge q' \wedge p) \vee (r \wedge q' \wedge p')$$

Of this we notice that the first bracket is equivalent to the third, the fourth to the eighth and the fifth to the seventh. The reiterated brackets can be omitted, (1.28) and (1.3), and the given proposition assumes the full disjunctive normal form:

$$(q \wedge p' \wedge r) \vee (q \wedge p' \wedge r') \vee (r \wedge p' \wedge q') \vee (p \wedge q' \wedge r) \vee (p \wedge q' \wedge r')$$

From this it can be inferred that the given proposition has the value 'true' when any one of the conjunctions has that value and, by an argument the exact dual of the one already given, only then. The appropriate combinations are:

| | |
|---|---|
| $p', q, r$ | all 'true' |
| $p', q, r'$ | all 'true' |
| $p', q', r$ | all 'true' |
| $p, q', r$ | all 'true' |
| $p, q', r'$ | all 'true' |

and these are of course just the combinations unaccounted for in Example 6.

*Example 8*

*Peter wants to go camping but will only go if Quentin goes; Quentin is willing to go with Peter if Roger is included; Roger will only go with one other person at most. Who can go?*

Let $p$ stand for 'Peter goes camping'; $q$ for 'Quentin goes camping'; $r$ for 'Roger goes camping'. The conjunction of all the conditions is:

$$(p' \vee q) \wedge ((p \wedge q)' \vee r) \wedge (r' \vee p' \vee q')$$

In full conjunctive normal form this is:

$$(p' \vee q \vee r) \wedge (p' \vee q \vee r') \wedge (p' \vee q' \vee r) \wedge (p' \vee q' \vee r')$$

which has the value 'true' if and only if $p'$ has that value.

Peter cannot go camping. Quentin and Roger can go together or separately or not at all.

*Example 9*

*A light is to be operated by three switches, each of which can occupy two positions – 'up' and 'down'. The changing of any switch must change the condition of the light – either from 'on' to 'off' or from 'off' to 'on', whatever the position of the other switches. Devise a simple circuit, using 'not-equivalent' gates.*

Suppose the switches are $P$, $Q$ and $R$ and $p$ stands for '$P$ is down', etc. Let $s$ stand for 'light is on'. The given conditions amount to:

$$\begin{aligned} s &\equiv ((p \wedge q \wedge r) \vee (p' \wedge q' \wedge r) \vee (p' \wedge q \wedge r') \vee (p \wedge q' \wedge r')) \\ &\equiv (((p \wedge q) \vee (p' \wedge q')) \wedge r) \vee (((p \wedge q') \vee (p' \wedge q)) \wedge r') \\ &\equiv ((p \equiv q) \wedge r) \vee ((p \equiv q)' \wedge r') \\ &\equiv ((p \equiv q) \equiv r) \end{aligned}$$

The required circuit is that shown in Figure 1.1, in which the effect of a 'not equivalent' gate, symbolised by $\not\equiv$ is that a voltage is produced on output if one or other input maintains a voltage but not both.

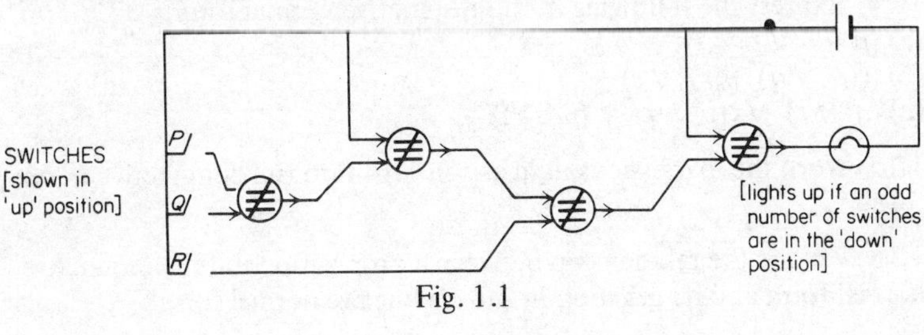


Fig. 1.1

*Example 10*

*Let x, y, z be ordinary numbers and let p stand for '$x > y$' (x is strictly greater than y); q for '$y > z$'; and r for '$z > x$' and let the 'truth' of* $(p \vee q \vee r) \wedge (p' \vee q' \vee r')$ *be assumed. It is required to construct a proposition that is 'false' if y lies between x and z, but 'true' otherwise.*

The conditions which secure the required falsehood are:

| | |
|---|---|
| $p', q', r$ | all 'false' |
| $p, q, r'$ | all 'false' |

So, in full conjunctive normal form the required proposition is:

$$(p' \vee q' \vee r) \wedge (p \vee q \vee r')$$

Notice that if we had not assumed that $p, q, r$ cannot all be simultaneously 'true' or simultaneously 'false' the required proposition would have been:

$$(p' \vee q' \vee r) \wedge (p' \vee q' \vee r') \wedge (p \vee q \vee r) \wedge (p \vee q \vee r')$$

---

**Exercises**

1.11 Give the solution to the problem of the snake in the box in full conjunctive normal form.

1.12 Write out in full the dual of the principle enunciated in Example 6 (p.16).

1.13 Express in full conjunctive normal form propositions having four constituents – $p, q, r, s$ – that are respectively 'false' when:
(a) $p, q, r$ are all 'true'; but 'true' otherwise
(b) $p, q, r$ are all 'false'; but 'true' otherwise
(c) $p, q$ are both 'true' and $r, s$ both 'false' or vice versa; but true otherwise.

1.14 Express the following in full disjunctive normal form:
(a) $(p \vee q \vee r) \wedge (p' \vee q')$
(b) $(((p' \vee q)' \vee r)' \vee s)$
(c) $(q' \vee r)' \vee ((p \vee q)' \vee (p \vee r))$

1.15 Work the expressions in question 1.14 into full conjunctive normal form.

1.16 What is the relationship between a proposition in full conjunctive normal form and its negation in full disjunctive normal form?

1.17 $A$ will form an alliance with $B$ if $C$ is included, but not otherwise; $B$ will form an alliance with $A$ or $C$, but not with both. Can $C$ form an alliance with $A$ but not with $B$?

1.18 A car's interior light is governed by a switch which may be in one of three positions. If in position 1, the light comes on if and only if the driver's door is open; if in position 2 the light is on whether the door is open or not; if in position 3 the light is off whether the door is open or not. Draw a circuit using 'nor' gates.

1.19 Examine the validity of the following argument. If $A$ is $B$'s mother then $C$ is not $B$'s father. But if $A$ is not $B$'s mother, then $A$ is married to $C$. Therefore it is not the case both that $C$ is $B$'s father and that $A$ is married to $C$. (London)

1.20 'If $p$ then $q$ else $r$' is defined as $(p' \vee q) \wedge (p \vee r)$. Using $n$ for a contradiction, $(s \wedge s')$, and $u$ for a tautology, $(s \vee s')$, state which of the following are tautologies:
(a) 'if $p$ then $q$ else $u$' $\equiv (q \Rightarrow p)$
(b) 'if $p$ then $q$ else $n$' $\equiv (p \vee q)$
(c) 'if $p$ then $u$ else $q$' $\equiv (p \wedge q)$ (Edinburgh)

---

## 1.6 Axioms

This section may be omitted at first reading without loss of continuity.

### (a) *The need of axioms*

In accepting the 'justification' offered in Example 6 (p.16) of the principle which states that any proposition can be expressed in full conjunctive normal form you may have felt a little uneasy; in fact you ought to have done so! For nothing more was demonstrated than that in a particular case – of three constituent propositions – a rule could be found for finding the full conjunctive normal form where the truth-table of the proposition in every case was known. And the effectiveness of that rule depended on our being able to show that in that full conjunctive normal form no two of the propositions conjoined can be 'false' at the same time.

From the point of view of strict logic this 'demonstration' suffers from two serious defects. In the first place we are left to see for ourselves that the same 'demonstration' can be extended to the case of four, five or any number of constituent propositions (a concept we have not hitherto made use of). In other words an appeal is made – though not explicitly – to the principle of **mathematical induction** (Chapter 4).

And in the second place the demonstration that no two of the disjunctions conjoined could be 'false' at the same time depended, in the last resort, upon writing down every possible combination of truth-values, finding the truth-values of all the possible disjunctions in each case and showing that in no single case were two disjunctions simultaneously 'false' – a purely mechanical procedure.

In other words we have so far established no general demonstration of the principle in question and upon the basis of rules so far laid down (about negating and conjoining propositions and the corresponding truth-values to be assigned to the result) no such general demonstration is possible. What alternative assertions (or **axioms**) could we use as a basis for such general principles as this?

(b) *Axioms of Russell and Whitehead*

It was partly with the idea of answering such questions that *Principia Mathematica*[10] was written. The widespread use of modern logic dates from its publication and the preliminary conclusion of its authors was that five primary axioms would be needed. These are:

$$(p \vee p) \Rightarrow p \tag{1.64}$$
$$q \Rightarrow (p \vee q) \tag{1.65}$$
$$(q \Rightarrow r) \Rightarrow ((p \vee q) \Rightarrow (p \vee r)) \tag{1.66}$$
$$(p \vee (q \vee r)) \Rightarrow (q \vee (p \vee r)) \tag{1.67}$$
$$(p \vee q) \Rightarrow (q \vee p) \tag{1.68}$$

The fourth axiom (1.67) can in fact be obtained from the others, and no use is made of it in what follows.

In addition to the axioms themselves, two **rules** are required for their manipulation.

The first is the **rule of substitution** which states: any proposition may be substituted for $p$, $q$, $r$. This rule inheres in the form of the axioms themselves, since we are to understand that they are asserted of all propositions whatsoever.

The second rule is the **rule of inference**[11] (RI), which states: anything implied by a true proposition is true. This we take to mean that when $p$ and $p \Rightarrow q$ are asserted $q$, may be asserted also. This corresponds to the rule of valid inference enunciated in section **1.4**(a), but is not identical with it, since it does not make use of the notion of tautology and allows inference from a one-premiss implication only. The corresponding rule for two-premiss implication can however be obtained by applying this rule to the axioms (see 'The laws of conjunction', page 26).

Apart from what is asserted in the axioms and permitted in the rules, nothing is said of propositions as such, except that every proposition can be negated to yield another proposition and any two can be alternated to yield a third.[12] Negation is an undefined operation but material implication is defined by:

$$(p \Rightarrow q) \equiv (p' \vee q) \quad \text{Df} \tag{1.69}$$

The addendum 'Df' in association with the sign of equivalence means that the symbol on the left may be substituted for that on the right wherever the latter occurs.

This axiomatic approach to logic does not then require us to assume that every proposition must be 'true' or 'false'; indeed these concepts play no part in its development. The assertions of $(p \vee p')$ and of $(p \wedge p')'$ follow from the axioms themselves. And in this way the need to make more statements about propositions is kept to a minimum.

Why, you may be tempted to ask, are these five axioms, (1.64) – (1.68), chosen in particular? Could we not manage with less? Would not another set have done equally well? The answer to both these questions is 'yes'.[13] These axioms are chosen partly because of their 'obviousness'. They assert qualities which we should expect propositions in general to display. Furthermore they exhibit two out of the three most important characteristics sought in an axiom-set, namely **consistency**, **independence** and **completeness**. A few remarks are appropriate to each of these.

(c) *Consistency*

It has been pointed out that although the five axioms make use of the idea of negation (this being involved in the definition of implication) they make no use of the notions of 'truth' and 'falsehood'. They are indeed no more than an arrangement of symbols on a piece of paper and as such have no more meaning than we choose to give them. So, if we can show that for some meaning of the word 'true' the substitution for *p, q* and *r* of any propositions whatever yields axioms that are 'true', and if, further, we can show that any manipulation of 'true' axioms according to the rules (of substitution and inference) yields only 'true' assertions, then the possibility of any 'false' assertion being brought about by such manipulation is excluded at the outset; and the axioms are said to be **consistent**.

But we can show both these things precisely by the use of truth-tables. It is a simple exercise to show by this means that (1.64)–(1.68) are indeed tautologies. It is equally plain from an argument analogous to that of section **1.4**(a) that if $t$ and $t \Rightarrow s$ are tautologies then $s$ also must be a tautology (to suppose otherwise involves a contradiction).[14] So, if we use only the rules of substitution and inference on these five axioms we obtain only 'true' assertions as a result. Furthermore since on this understanding of 'truth' every proposition is either 'true' or 'false' and no proposition is both no contradiction can arise.

(d) *Independence*

Another desirable characteristic in an axiom-set, though in this case not a necessary one, is independence. By asserting of any such set that each axiom is **independent** of the others we mean to say that by no manipulation of the others according to the rules can that axiom be reached. This is not in fact true of the set under consideration. The fourth (1.67) can be inferred from the others. However it is the case that each of the remaining four is independent of the other three. The proof, for which you are referred to more advanced texts, consists in finding a meaning for the symbols and for the word 'true' under which three of these four axioms are 'true' while the fourth is not.

(e) *Completeness*

This is the most desirable characteristic of an axiom-set and the most difficult to secure. It arises in connection with the question of **provability**. The logic developed in sections 1.2–1.5 has the important property that any theorem in it is **decidable**. That is to say that of any complex propositions, however many constituent propositions it may involve, we can state definitely whether it is a tautology or not. All we need to do is to construct a truth-table incorporating every possible combination of truth-values in the constituent propositions and determine by means of our truth-table rules whether or not the complex proposition is 'true' in every case.

What we should also like to be able to show is that to every tautology in the first system there corresponds a **provable assertion** in the second, i.e., *an assertion that can be reached by means of an argument of which the only premisses are one or more of the five axioms.* When this can be shown the axiom-set is said to be **complete**. In fact the axiom-set of Russell and Whitehead is complete, though a demonstration of that fact is not easy.[16]

All we can hope to show here is that at any rate the basic tautologies of sections **1.2–1.5** are provable. That those proved are in turn sufficient for the proof of those remaining you may (if you have sufficient patience) show for yourself.

In the case of the first three proofs, every substitution made will be given in full as also a reference to the axioms or other earlier assertions referred to when the Rule of Inference (RI) is employed. After that in general only references will be given and you will be left to discover for yourself what substitutions are required.

(i) Law of successive implication

$$(q \Rightarrow r) \Rightarrow ((p \Rightarrow q) \Rightarrow (p \Rightarrow r)) \qquad (1.70)$$

(1.66: $p'$ for $p$)

(ii) Laws of the excluded middle

$$p' \vee p \quad \text{and} \quad p \vee p'$$

$$p \Rightarrow (p \vee p) \tag{1.72}$$

(1.65: $p$ for $q$)

$$(p \vee p) \Rightarrow p \tag{1.73}$$

(1.64)

$$((p \vee p) \Rightarrow p) \Rightarrow ((p \Rightarrow (p \vee p)) \Rightarrow (p \Rightarrow p)) \tag{1.74}$$

(1.66: $(p \vee p)$ for $q$, $p$ for $r$, $p'$ for $p$)

$$((p \vee p) \Rightarrow p) \Rightarrow (p \Rightarrow p) \tag{1.75}$$

(RI) (1.73) and (1.74)

$$p \Rightarrow p \tag{1.76}$$

(RI) (1.72) and (1.75)

$$p' \vee p \tag{1.77}$$

Definition (1.69: $p$ for $q$)

$$(p' \vee p) \Rightarrow (p \vee p') \tag{1.78}$$

(1.68: $p'$ for $p$, $p$ for $q$)

$$p \vee p' \tag{1.79}$$

(RI) (1.77) and (1.78)

(1.77) and (1.79) are the assertions to be proved.

(iii) Laws of double negation

$$p \Rightarrow (p')' \quad \text{and} \quad (p')' \Rightarrow p$$

$$p'' \equiv (p')' \quad \text{Df} \tag{1.81}$$

$$p \Rightarrow p'' \tag{1.82}$$

(1.79: $p'$ for $p$) and (1.81)

$$p'' \Rightarrow (p'')'' \tag{1.83}$$

(1.82: $p''$ for $p$)

$$p' \vee (p'')'' \tag{1.84}$$

(1.69: $p''$ for $q$) and (1.81)

$$(p' \vee (p'')'') \Rightarrow ((p'')'' \vee p') \tag{1.85}$$

(1.68: $p''$ for $p$, $(p'')''$ for $q$)

$$(p'')'' \vee p' \tag{1.86}$$

(RI) (1.84) and (1.85)

$$(p')'' \vee p \tag{1.87}$$

(1.86: $p$ for $p'$)

$$(p')' \Rightarrow p \tag{1.88}$$

(1.69: $p'$ for $p$, $p$ for $q$)

(1.82) and (1.88) are the assertions to be proved.

(iv) Laws of negative implication

$$(p \Rightarrow q) \Rightarrow (q' \Rightarrow p')$$

and $$(p' \Rightarrow q') \Rightarrow (q \Rightarrow p)$$

$$(q \Rightarrow q'') \Rightarrow ((p' \vee q) \Rightarrow (p' \vee q'')) \qquad (1.90)$$

(1.66: $q''$ for $r$)

$$(p' \vee q) \Rightarrow (p' \vee q'') \qquad (1.91)$$

(RI) (1.82) and (1.90)

$$(p' \vee q'') \Rightarrow (q'' \vee p') \qquad (1.92)$$

(1.68: $p'$ for $p$, $q''$ for $q$)

$$(p' \vee q) \Rightarrow (q'' \vee p') \qquad (1.93)$$

(RI) (1.70) and (1.92)

$$(p \Rightarrow q) \Rightarrow (q' \Rightarrow p') \qquad (1.94)$$

(1.69: $q'$ for $p$, $p'$ for $q$) and (1.93)

We have not yet justified the use of successive implication within a subsidiary expression, so we cannot simply replace $p$ by $p'$, $q$ by $q'$ and use (iii) to obtain the converse. Instead:

$$(p'' \Rightarrow p) \Rightarrow ((q' \vee p'') \Rightarrow (q' \vee p) \qquad (1.95)$$

(1.66: $p''$ for $q$, $p$ for $r$, $q'$ for $p$)

$$p'' \Rightarrow p \qquad (1.96)$$

(1.88)

$$(q' \vee p'') \Rightarrow (q' \vee p) \qquad (1.97)$$

(RI) (1.95) and (1.96)

$$(p'' \vee q') \Rightarrow (q' \vee p'') \qquad (1.98)$$

(1.68: $p''$ for $p$)

$$((q' \vee p'') \Rightarrow (q' \vee p)) \Rightarrow (((p'' \vee q') \Rightarrow (q' \vee p'')) \Rightarrow ((p'' \vee q') \Rightarrow (q' \vee p))) \qquad (1.99)$$

(1.70: $(p'' \vee q')$ for $p$, $(q' \vee p'')$ for $q$, $(q' \vee p)$ for $r$)

$$((p'' \vee q') \Rightarrow (q' \vee p'')) \Rightarrow ((p'' \vee q') \Rightarrow (q' \vee p)) \qquad (1.100)$$

(RI) (1.97) and (1.99)

$$(p'' \vee q') \Rightarrow (q' \vee p) \qquad (1.101)$$

(RI) (1.98) and (1.100)

$$(p' \Rightarrow q') \Rightarrow (q \Rightarrow p) \qquad (1.102)$$

(1.69: $p'$ for $p$, $q'$ for $q$) and (1.101)

(1.94) and (1.102) are the assertions to be proved.

(v) Laws of conjunction

$$(p \wedge q) \Rightarrow q; \quad (p \wedge q) \Rightarrow p$$

$$(p \wedge q) \equiv (p' \vee q')' \quad \text{Df} \qquad (1.104)$$

$$(p \wedge q) \equiv (p' \vee q')' \tag{1.105}$$

Df

$$q' \Rightarrow (p' \vee q') \tag{1.106}$$

(1.65: $q'$ for $q$, $p'$ for $p$)

$$(q' \Rightarrow (p' \vee q')) \Rightarrow ((p' \vee q')' \Rightarrow q'') \tag{1.107}$$

(1.94: $q'$ for $p$, $(p' \vee q')$ for $q$)

$$(p' \vee q')' \Rightarrow q'' \tag{1.108}$$

(RI) (1.106) and (1.107)

$$q'' \Rightarrow q \tag{1.109}$$

(1.88: $q$ for $p$)

$$(q'' \Rightarrow q) \Rightarrow (((p' \vee q') \Rightarrow q'') \Rightarrow ((p' \vee q') \Rightarrow q))) \tag{1.110}$$

(1.70: $(p' \vee q')'$ for $p$, $q''$ for $q$, $q$ for $r$)

$$((p' \vee q')' \Rightarrow q'') \Rightarrow ((p' \vee q')' \Rightarrow q) \tag{1.111}$$

(RI) (1.109) and (1.110)

$$(p' \vee q')' \Rightarrow q \tag{1.112}$$

(RI) (1.109) and (1.111)

$$(p' \vee q') \Rightarrow (q' \vee p') \tag{1.113}$$

(1.68: $p'$ for $p$, $q'$ for $q$)

$$((p' \vee q') \Rightarrow (q' \vee p')) \Rightarrow ((q' \vee p')' \Rightarrow (p' \vee q')') \tag{1.114}$$

(1.94: $(p' \vee q')$ for $p$, $(q' \vee p')$ for $q$)

$$(q' \vee p')' \Rightarrow (p' \vee q')' \tag{1.115}$$

(RI) (1.113) and (1.114)

$$((p' \vee q')' \Rightarrow q) \Rightarrow (((q' \vee p')' \Rightarrow (p' \vee q')') \Rightarrow ((q' \vee p')' \Rightarrow q)) \tag{1.116}$$

(1.70: $(q' \vee p')'$ for $p$ $(p' \vee q')'$ for $q$, $q$ for $r$)

$$((q' \vee p')' \Rightarrow (p' \vee q')') \Rightarrow (((q' \vee p')' \Rightarrow q) \tag{1.117}$$

(RI) (1.112) and (1.116)

$$(q' \vee p')' \Rightarrow q \tag{1.118}$$

(RI) (1.115) and (1.117)

$$(p \wedge q) \Rightarrow q \tag{1.119}$$

(RI) (1.105) and (1.112)

$$(q \wedge p) \Rightarrow q \tag{1.120}$$

Df (1.105) and (1.118)

$$(p \wedge q) \Rightarrow p \tag{1.121}$$

(1.120: $p$ for $q$, $q$ for $p$)

Furthermore, from $p$, $q$ we can infer $q \wedge p$ as follows:

$$p \tag{1.122}$$

Premiss

$$p \Rightarrow (q' \vee p) \qquad (1.123)$$

(1.65: $p$ for $q$, $q'$ for $p$)

$$q' \vee p \qquad (1.124)$$

(RI) (1.122) and (1.123)

$$q \Rightarrow p \qquad (1.125)$$

Df (1.69) and (1.124)

$$p' \Rightarrow q' \qquad (1.126)$$

(RI) (1.94: $p$ for $q$, $q$ for $p$) and (1.125)

$$(p' \Rightarrow q') \Rightarrow ((q' \vee p') \Rightarrow (q' \vee q')) \qquad (1.127)$$

(1.66: $p'$ for $q$, $q'$ for $r$, $q'$ for $p$)

$$(q' \vee p') \Rightarrow ((q' \vee q') \qquad (1.128)$$

(RI) (1.126) and (1.127)

$$(q' \vee q') \Rightarrow q' \qquad (1.129)$$

(1.64: $q'$ for $p$)

$$((q' \vee q') \Rightarrow q') \Rightarrow (((q' \vee p') \Rightarrow (q' \vee q')) \Rightarrow ((q' \vee p') \Rightarrow q')) \qquad (1.130)$$

(1.70: $(q' \vee p')$ for $p$, $(q' \vee q')$ for $q$, $q'$ for $r$)

$$((q' \vee p') \Rightarrow (q' \vee q') \Rightarrow ((q' \vee p') \Rightarrow q') \qquad (1.131)$$

(RI) (1.129) and (1.130)

$$(q' \vee p') \Rightarrow q' \qquad (1.132)$$

(RI) (1.128) and (1.131)

$$((q' \vee p') \Rightarrow q') \Rightarrow (q'' \Rightarrow (q' \vee p')') \qquad (1.133)$$

(1.94: $(q' \vee p')$ for $p$, $q'$ for $q$)

$$q'' \Rightarrow (q' \vee p')' \qquad (1.134)$$

(RI) (1.132) and (1.133)

$$q \Rightarrow q'' \qquad (1.135)$$

(1.88: $q$ for $p$)

$$(q'' \Rightarrow (q' \vee p')') \Rightarrow ((q \Rightarrow q'') \Rightarrow (q \Rightarrow (q' \vee p')') \qquad (1.136)$$

(1.70: $q$ for $p$, $q''$ for $q$, $(q' \vee p')$ for $r$)

$$(q \Rightarrow q'') \Rightarrow (q \Rightarrow (q' \vee p')') \qquad (1.137)$$

(RI) (1.134) and (1.136)

$$q \Rightarrow (q' \vee p')' \qquad (1.138)$$

(RI) (1.135) and (1.137)

$$q \Rightarrow (q \wedge p) \qquad (1.139)$$

Df (1.104: $q$ for $p$, $p$ for $q$) and (1.138)

$$q \qquad (1.140)$$

Premiss

$$q \wedge p \qquad (1.141)$$

(RI) (1.139) and (1.140)

Plainly, by interchanging $p$ and $q$ throughout the argument we could equally have inferred $p \wedge q$. This important result allows us when the conjunction of two propositions is asserted to infer either of them separately; and when both are asserted to infer their conjunction.[17]

(vi) Substitution of equivalents

From $p \Leftrightarrow q$, $q \Leftrightarrow r$ we can infer $p \Leftrightarrow r$

Df

$$(p \Leftrightarrow q) \equiv (p \Rightarrow q) \wedge (q \Rightarrow p) \qquad (1.143)$$

$$(p \Rightarrow q) \wedge (q \Rightarrow p) \qquad (1.144)$$

Premiss Df

$$((p \Rightarrow q) \wedge (q \Rightarrow p)) \Rightarrow (p \Rightarrow q) \qquad (1.145)$$

(1.121) $(p \Rightarrow q)$ for $p$, $(q \Rightarrow p)$ for $q$)

$$p \Rightarrow q \qquad (1.146)$$

(RI) (1.144) and (1.145)

$$(q \Rightarrow r) \wedge (r \Rightarrow q) \qquad (1.147)$$

Premiss Df

$$((q \Rightarrow r) \wedge (r \Rightarrow q)) \wedge (q \Rightarrow r) \qquad (1.148)$$

(1.121) $(q \Rightarrow r)$ for $p$, $(r \Rightarrow q)$ for $q$)

$$q \Rightarrow r \qquad (1.149)$$

(RI) (1.147) and (1.148)

$$(q \Rightarrow r) \Rightarrow ((p \Rightarrow q) \Rightarrow (p \Rightarrow r)) \qquad (1.150)$$

(1.70)

$$(p \Rightarrow q) \Rightarrow (p \Rightarrow r) \qquad (1.151)$$

(RI) (1.149) and (1.150)

$$p \Rightarrow r \qquad (1.152)$$

(RI) (1.146) and (1.151)[18]

Interchanging $p$ and $r$ in (1.144)–(1.152) leads to:

$$r \Rightarrow p \qquad (1.153)$$

$$(p \Rightarrow r) \wedge (r \Rightarrow p) \qquad (1.154)$$

(1.121); $(p \Rightarrow r)$ for $p$ $(r \Rightarrow p)$ for $q$) (1.152) and (1.153)

$$p \Leftrightarrow r \qquad (1.155)$$

(1.154) and (1.142)

It is evident from this that in the course of an argument any proposition may be substituted for an equivalent proposition on either side of a sign of equivalence. The ensuing arguments justify us in substituting equivalent propositions for one another anywhere.

From $t \Leftrightarrow u$, $s \Leftrightarrow w$ we infer $(t \vee s) \Leftrightarrow (u \vee w)$

$$s \Leftrightarrow w \tag{1.157}$$
Premiss

$$s \Rightarrow w \tag{1.158}$$
(1.121) and (1.157)

$$(s \Rightarrow w) \Rightarrow ((p \vee s) \Rightarrow (p \vee w)) \tag{1.159}$$
(1.66)

$$(p \vee s) \Rightarrow (p \vee w) \tag{1.160}$$
(1.158) and (1.159)

$$w \Rightarrow s \tag{1.161}$$
(1.121) and (1.157)

$$(w \Rightarrow s) \Rightarrow ((p \vee w) \Rightarrow (p \vee s)) \tag{1.162}$$
(1.66)

$$(p \vee w) \Rightarrow (p \vee s) \tag{1.163}$$
(1.161) and (1.162)

$$(p \vee s) \Leftrightarrow (p \vee w) \tag{1.164}$$
(1.122), (1.160) and (1.163)

$$(s \vee p) \Rightarrow (p \vee s) \tag{1.165}$$
(1.68)

$$(p \vee s) \Rightarrow (s \vee p) \tag{1.166}$$
(1.68)

$$(s \vee p) \Leftrightarrow (p \vee s) \tag{1.167}$$
(1.143), (1.165) and (1.166)

Using the obvious notation justified by (1.143), we infer:

$$(s \vee p) \Leftrightarrow (p \vee s) \Leftrightarrow (p \vee w) \Leftrightarrow (w \vee p) \tag{1.168}$$
(1.164) and (1.167)

and similarly

$$(t \vee q) \Leftrightarrow (q \vee t) \Leftrightarrow (q \vee u) \Leftrightarrow (u \vee q) \tag{1.169}$$

The substitution of $t$ for $p$ in (1.164) and $w$ for $q$ in (1.165) yields the required inference.

From $t \Leftrightarrow u$, $s \Leftrightarrow w$ we infer $t' \Leftrightarrow u'$, $s' \Leftrightarrow w'$ (1.170)

The argument follows the same lines as that of the foregoing with an appeal to (1.94) in place of (1.66).

From $t \Leftrightarrow u$, $s \Leftrightarrow w$ we infer $(t \wedge s) \Leftrightarrow (u \wedge w)$ (1.171)

This follows immediately from what precedes, (1.120) and (1.121) and the laws of conjunction.

The right to substitute equivalents, each for the other, anywhere within a complex proposition in the course of an argument justifies us in dropping ⇔ in favour of ≡, the symbol for equivalence used in section (1.3) et seq., since equivalence confers exactly the same rights as definition. From the point of view of manipulation equivalent propositions are identical.

There remain only a few equivalences to be established that are in frequent use and these are left as exercises for the reader. They are listed here, for convenience.

(vii) Laws of identity, idempotence and absorption

$$p \equiv p$$
$$p \equiv (p \vee p)$$
$$p \equiv (p \wedge p)$$
$$p \equiv (p \vee (p \wedge q))$$
$$p \equiv (p \wedge (p \vee q))$$

(viii) Commutative, associative and distributive laws

$$(p \vee q) \equiv (q \vee p)$$
$$(p \wedge q) \equiv (q \wedge p)$$
$$((p \vee q) \vee r) \equiv ((p \vee (q \vee r))$$
$$((p \wedge q) \wedge r) \equiv ((p \wedge (q \wedge r))$$
$$(p \wedge (q \vee r)) \equiv ((p \wedge q) \vee (p \wedge r))$$
$$(p \vee (q \wedge r)) \equiv ((p \vee q) \wedge (p \vee r))$$

These form three self-dual pairs.

(ix) Conjunction and alternation with tautologies and contradictions

$$(q \wedge q') \equiv (p \wedge p')$$
$$(q \vee q') \equiv (p \vee p')$$
$$q \vee (p \wedge p') \equiv q$$
$$q \wedge (p \vee p') \equiv q$$
$$q \wedge (p \wedge p') \equiv (p \wedge p')$$
$$q \vee (p \vee p') \equiv (p \vee p')$$

The first two equivalences here justify us in writing $u$ for any proposition equivalent to $(p \vee p')$ (a tautology) and $n$ for any proposition equivalent to $(p \wedge p')$ (a contradiction). The last four again form self-dual pairs.

The results of (vi)–(ix) form the basis of a complete algebra of equivalences, similar in many respects to the algebra of the natural numbers whose fundamental laws we shall consider in chapter 4. Occasionally supplemented by results obtained by the truth-table method (which can also be reached by inference from the axioms as was pointed out in section **1.6**(e)) they suffice for the development of a complete calculus of propositions, sometimes called the **sentential calculus.**

---

**Exercises**

1.21 Verify by truth-tables that (1.64)–(1.68) are tautologous in the sense of section **1.3**(c).

1.22 Obtain the relations in parts (vii) to (ix) by substitutions in and inferences from assertions already made.[19]

1.23 Establish the equivalent of an equivalence:

$$(p \equiv q) \equiv ((p \wedge q) \vee (p' \wedge q'))$$

1.24 Prove the following by inferences from the axioms or assertions derived from them:

(a) $p' \Rightarrow (p \Rightarrow q)$
(b) $((p \wedge q) \vee (q \wedge r) \vee (r \wedge p)) \equiv ((p \vee q) \wedge (q \vee r) \wedge (r \vee p))$
(c) $(p \Rightarrow q) \Rightarrow (((p \vee (q \wedge r)) \equiv (q \wedge (p \vee r)))$

**Miscellaneous exercises**

1.25 Demonstrate the validity of the following arguments, using truth-tables:

(a)
$$\begin{array}{l} (p \vee q) \Rightarrow r \\ r' \\ \hline p' \end{array}$$

(b)
$$\begin{array}{l} (p \wedge q) \vee r \\ q \Rightarrow r' \\ q \\ \hline p \end{array}$$

(c)
$$\begin{array}{l} (p \wedge q') \Rightarrow r \\ p' \Rightarrow q \\ q' \\ \hline r \end{array}$$

(London)

1.26 Put the following arguments in symbolic form and test their validity:

(a) Unless there is a breakdown somewhere the train will be on time, in which case you will get to work in plenty of time and I won't be late for my appointment. Since there won't be a breakdown I will not be late for my appointment.

(b) The documents are either in Russian or Bulgarian. Russian and Bulgarian are both Slavonic languages. The only man who has had access to these documents does not understand Slavonic languages; so either he has not read the documents or if he has he has not understood them.

1.27 Simplify the following proposition:

$$(p \wedge q' \wedge r' \wedge s') \vee (t \wedge p' \wedge r) \vee (t \wedge p' \wedge q \wedge r') \vee (t' \wedge p \wedge r) \vee (t' \wedge p \wedge r' \wedge s) \vee (t' \wedge p' \wedge q) \vee (t' \wedge p \wedge q' \wedge s').$$

1.28 The following table expresses the relation obtaining between two outputs *a*, *b* and four inputs *w*, *x*, *y*, *z* in a network. No combination of inputs can occur other than those shown. Design a suitable network, using 'and', 'or' and 'not' gates.

| *w* | *x* | *y* | *z* | *a* | *b* |
|---|---|---|---|---|---|
| 0 | 0 | 1 | 0 | 1 | 1 |
| 0 | 0 | 1 | 1 | 0 | 1 |
| 0 | 1 | 0 | 0 | 0 | 0 |
| 0 | 1 | 0 | 1 | 0 | 0 |
| 0 | 1 | 1 | 0 | 1 | 0 |
| 0 | 1 | 1 | 1 | 1 | 1 |
| 1 | 0 | 0 | 1 | 1 | 0 |
| 1 | 0 | 1 | 0 | 1 | 1 |
| 1 | 1 | 0 | 1 | 0 | 0 |
| 1 | 1 | 1 | 0 | 1 | 1 |
| 1 | 1 | 1 | 1 | 1 | 1 |

(Courtesy of the British Computer Society)

1.29

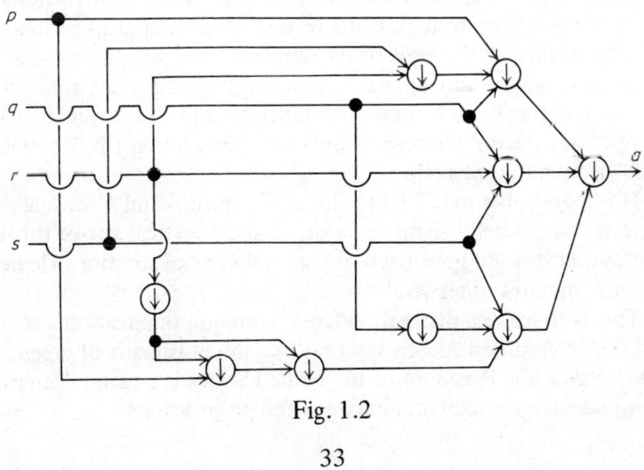


Fig. 1.2

The network in figure 1.2 employs only 'nand' gates; that is, voltages are produced on output except when voltages are maintained on all inputs simultaneously. Design an equivalent and simpler network, also using 'nand' gates only. (Courtesy of the British Computer Society).

1.30 Five witnesses to an event *A, B, C, D* and *E* give conflicting accounts of it and of one another's veracity.

*A* says: '*C* and *D* are lying'
*B* says: '*A* and *E* are lying'
*C* says: '*B* and *D* are lying'
*D* says: '*C* and *E* are lying'
*E* says: '*A* and *B* are lying'

Assuming that none of these statements is self-contradictory, which, if any, can be shown to be false?

---

**Notes on Preface and chapter 1**

1. e.g., Courant and Robbins, *What is Mathematics?* (Oxford).
2. A notable exception is Archimedes who seems to have been aware of the fact that his assumptions were not self-evident. For his use of logic to prove an even more fundamental physical principle see Cohen and Nagel, *Introduction to Logic and Scientific Method* (Routledge and Kegan Paul), pp. 407–413.
3. The term 'statement' is reserved for statements *about* propositions. Thus the question is sidestepped whether our definition of a proposition is itself a proposition!
4. It is not necessary to know what the truth-value of a proposition is in order to know that it has one. The truth-value of 'George Joseph Smith murdered three women' may be in dispute. The truth-value of 'today is Thursday' varies from time to time. Nevertheless each proposition is either true or false at the time of its assertion.
5. At first sight it may appear that an equivalence is not a proposition at all, but a statement about two propositions, but this is not so. This point becomes clearer when we distinguish – in section **1.6** – between a proposition and its assertion.
6. It is regrettable that ↑ has to be used for 'nor' and ↓ for 'nand'; a vertical stroke generally signifies negation and a vertical stroke through the 'or' symbol would have been the natural symbol for 'nor'. However, convention dictates otherwise!
7. This is to make a distinction from 'common implication'. We should not in common speech say that 'the moon is made of green cheese' *implies* 'the President of the United States is a pauper', there being no *necessary* connexion between the propositions.

8. The distinction here made between a proposition and its assertion belongs to the field of metalogic. It is sufficient here to point out that there clearly *is* a distinction. We may well wish to *consider* two or more propositions, for example, without in fact asserting them, and thus perhaps show that if they *were* simultaneously asserted a contradiction would ensue.

9. For this illustration I am indebted to Whitesitt, *Boolean Algebra and its Applications* (Addison-Wesley).

10. Whitehead and Russell, *Principia Mathematica* (Cambridge).

11. For the development of this approach to the calculus of propositions only one rule of inference is required, namely

$$\frac{\begin{array}{l} p \\ p \Rightarrow q \end{array}}{q}$$

The other rules given in section **1.4**(b) may all be developed from this. See, for example, Notes 17 and 18.

12. In *Principia Mathematica* conjunction is defined in terms of negation and alternation.

13. See Russell *Introduction to Mathematical Philosophy* (Cambridge) p. 152.

14. *t, s* here stand for particular combinations (such as (1.64)–(1.68)) of general propositions, *p, q, r.*

15. See Suppes, *Introduction to Logic* (Van Nostrand Reinhold), p. 64.

16. If of course we could assume the principle of the complete normal form, enunciated in section **1.5**(a), the completeness of the axiom-set would follow at once. For it is not difficult to show as a consequence of the axioms that from the assertion of any proposition we can infer any disjunction in which it occurs; and that from the assertion of all the propositions occurring in a conjunction we can infer the conjunction in which they occur. And since the axioms provide us with a means of asserting the equivalence of any complex proposition with its complete normal form it would be possible from this equivalence to determine whether it was tautologous or not. But it was precisely the lack of any rigorous demonstration of this principle that drove us to the axiomatic approach in the first place.

17. See section **1.4**(b), example (ii), page 14.

18. The inference of (1.152) from (1.146) and (1.149) by means of (1.144) and (1.151) is, after this point, made directly. It is in effect the syllogistic inference (see section **1.4**(a)(ii)) and (section **1.4**(b). Example (ii)).

19. They should be proved in the order in which they appear since some of the later assertions can be reached most naturally by inference from the earlier.

**Definitions used in chapter 1**

(1) $p'$ for $p \uparrow p$

(2) $p \vee q$ for $(p \uparrow q) \uparrow (p \uparrow q)$

(3) $p \Rightarrow q$ for $p' \vee q$

(4) $p \wedge q$ for $(p' \vee q')'$

(5) $p \equiv q$ for $(p \Rightarrow q) \wedge (q \Rightarrow p)$

(6) $p \wedge q \wedge r$ for $(p \wedge q) \wedge r$

(7) $p \vee q \vee r$ for $(p \vee q) \vee r$

# 2

# THE LOGIC OF SETS

## 2.1 Fact and hypothesis: Boyle's law

In the last chapter we examined the logic of the proposition that unequal bodies fall with equal accelerations. If in fact we do not know what that acceleration is, the proposition is only of limited use. In this chapter we will take a look at a premiss which yields more precise inferences.

Boyle's Law, in its simplest form, states that of any given quantity of gas the product of the pressure with the ratio of volume to absolute temperature is constant. This is expressed in the familiar formula:

$$\frac{P \times V}{T} = k$$

Such a general statement, covering all[1] possible variations of volume, pressure and temperature is called an **hypothesis**. How can we demonstrate that such a hypothesis is even approximately 'true'?

Let us assume that we have available a gas-tight cylinder whose volume we can, within certain limits, vary measurably and at will; and that instruments are incorporated for measuring the temperature and pressure of the gas within it.

The logical approach would then be: first, keep the temperature of the gas constant, vary the volume so as to give rise to a number of prescribed pressures and note the ratio of volume (measured in $cm^3$) to temperature (measured in K) in each case. Next, alter the temperature and repeat the experiment, noting the new ratios. And thus continue, making the necessary comparison of volume to temperature for each of the pressures prescribed.

After forming the product (that is, multiplying the pressure by the ratio of volume to temperature) in each case, we might tabulate the results as in Table 2.1.

Table 2.1.

Values[2] of $\frac{PV}{T}$ for 1 000 cm$^3$ of air at 273 K and atmospheric pressure (about $10^5$ Nm$^{-2}$)

| Pressure in Nm$^{-2}$ x $10^3$ | Temperature in Kelvins 273 | 289 | 373 |
|---|---|---|---|
| 100 | 371 | 371 | 371 |
| 1 000 | 369 | 370 | 372 |
| 1 500 | 368 | 369 | 372 |
| 2 000 | 367 | 369 | 372 |
| 2 500 | 366 | 368 | 372 |
| 3 000 | 365 | 367 | 372 |
| 3 500 | 365 | 367 | 372 |
| 4 000 | 364 | 366 | 372 |

This table conveys fact. It is therefore a proposition – and (within the limits afforded by the equipment and its user) a true one.

Plainly, however, it is not a simple proposition. If we take account only of the first column of figures, relating to measurements made at 273 K, this represents a conjunction of some eight propositions and even these are not simple since the quantities referred to in the formula $\frac{PV}{T}$ are not chosen at random.

The first of the conjoined propositions stated in full is:

'When the volume of air considered was 1 000 cm$^3$ and its temperature was 273 K, then the value of $\frac{PV}{T}$ was found to be 366.'

Nor would this proposition, conjoined with 23 others like it, completely represent the fact conveyed by the table, since such a conjunction would give no special significance to the fact that it was the same quantity of air, though occupying different volumes, which had been the subject of this experiment throughout. A completely logical enunciation of any hypothesis must make it clear what it is that is permitted to vary – here volume, temperature and pressure – and what – the quantity of air – is not. In fact in this case we get a rather more satisfactory hypothesis if we don't allow the volume of air to vary.

When we have finally decided upon the variable factors – for simplicity let us suppose that they are just temperature and pressure – we can include all the propositions conveyed by the first row of the table in the following general propositions:

'For a given quantity of air occupying a fixed volume of 1 000 $cm^3$ at 273 K and atmospheric pressure, the product of that volume with the ratio of its pressure to its absolute temperature lies between $363 \times 10^{-6}$ and $373 \times 10^{-6}$ J $K^{-1}$ at all temperatures.'

It is this last phrase 'at all temperatures' which is characteristic of a hypothesis or supposed set of facts; and it is the attempt thus to state a set of facts by means of one proposition which we are to scrutinise in this chapter.

In the first place it is worth noticing that the hypothesis set forth above goes beyond what the table proposes. The table itself only proposes that the quantity $\frac{PV}{T}$ lies between the suggested limits for all temperatures considered. Another, intermediate, temperature might have been considered, 300 K for example, and the result might have been quite different.

We could in fact have considered many other temperatures, though only a finite number of course, since we have only a limited time at our disposal. But the hypothesis itself does not require the number of cases to be restricted in any way[3] and herein lies its importance – it affords us unlimited opportunity of disproof.

Propositions like the one above are said to be quantified in respect of one (or more) variables – in this case of temperature alone. The phrase 'at all temperatures' or 'for all temperatures' is referred to as the **quantifier**, the whole proposition preceding as the **quantified proposition** and the temperature as the **variable of quantification**

## 2.2 Quantification

In a sense the idea of a quantified proposition is not new. The axiom $p \vee p \Rightarrow p$ (1.64) is asserted of all propositions $p$ and can thus be said to be quantified in respect of $p$. What is new in the proposition in section **2.1** is that part of the proposition in respect of which it is quantified – temperature – is not itself a proposition. We are familiar with the notion that a proposition, $p \vee q$ for example, may be analysed into parts $p$, $q$, which are themselves propositions. But up until now we have supposed (implicitly) that this process of analysis must come to an end with some 'atomic' proposition that could not be further analysed.

We are now faced with a question of some importance. Are there any 'atomic' propositions and, if not, into what components can the most fundamental of propositions be further analysed?

### (a) *Simple propositions*

The simplest analysis of anything would seem to be that which divides it into two parts, the parts themselves being related in some way. What exactly this relation may be in the case of a proposition is one which we choose to leave open at the moment, stipulating only that it serves to put the two parts in order so that we know which is which.

On this assumption we define a **simple proposition** as one of the form $x \in y$, where $x$, $y$ are the parts and $\in$ the relation between them so that the propositions $x \in y$ and $y \in x$ are regarded as distinct. The negation of this proposition we denote by $x \notin y$.

But what do $x$ and $y$ stand for? The most obvious answer and the one traditionally given is that $x$ is the **subject** of the proposition, i.e., it tells us what the proposition concerns; whereas $y$ is the **predicate** and yields such information concerning the subject as the proposition is intended to convey. In this case the meaning of the symbol $\in$, the **copula**, can best be rendered by *is* or *is a*.[4] A little practice will show that most propositions not involving comparisons can be expressed in this form.

For example:

*There is a tavern in that town* can be expressed as *that town* (subject) *is* (copula) *furnished with a tavern* (predicate)

*'Tis folly to be wise* can be expressed as *being wise* (subject) *is* (copula) *foolish* (predicate)

*A good breakfast promotes health* can be expressed as *a good breakfast* (subject) *is a* (copula) *health-promoting thing* (predicate)

However, *there are more fools than wise men* cannot easily be expressed in the subject–copula–predicate form; nor can *breakfast precedes lunch,* unless it is unnaturally forced into the form *breakfast is a lunch-preceding thing.*

(b) *The universal quantifier*

Having analysed a proposition – for the moment into subject and predicate – we can now quantify it.

Consider the first example cited above. If $x_0$ stands for *that town* and $Y$ for *furnished with a tavern,* the proposition *There is a tavern in that town* is represented by:

$$x_0 \in Y$$

Now suppose we want to quantify this, i.e., to form the proposition *There is a tavern in every town.* This amounts to saying that $x \in Y$ for all $x$ such that $x$ is a town. Or, so as to avoid the awkward *such that, x is a town implies x is furnished with a tavern, for all x.*

This is symbolised by:

$$(x)(x \in X \Rightarrow x \in Y) \tag{2.1}$$

where $X$ stands for that which is predicated of towns in general – the property of being a town.

The symbolism employed in (2.1) requires some explanation. The **universal quantifier** $(x)$, which is to be treated as one symbol, is used in reference to the whole of the bracket immediately following it but to that bracket only. In a complicated proposition $x$ may make several appearances not all of which are governed by the same quantifier or indeed by any quantifier at all.[5]

Within the bracket to which $(x)$ refers is a propositional function or **matrix**. Here it is $x \in X \Rightarrow x \in Y$; more generally it is denoted by $\phi x$. This matrix is not itself a proposition, but is capable of assuming that status when a definite significance – *Manchester* or *Mao Tse Tung* – is given to its argument,[6] $x$. The argument, for which a lower case letter is generally used, may be thought of as a place-keeper. It holds one or more places (two in this case) for occupation by another symbol of definite significance. It is therefore a dummy; provided the same symbol is used for the quantifier as is used for the argument, any lower case letter will do

$$(y)(y \in X \Rightarrow y \in Y)$$

has exactly the same significance as

$$(x)(x \in X \Rightarrow x \in Y)$$

and in general

$$(y)(\phi y) \equiv (x)(\phi x)$$

The expressions on each side of the sign of equivalence are said to be **closed** with respect to $x$ (or $y$); their significance depends only on $\phi$.

It may well be asked at this point: why introduce dummy arguments at all? Why not use a simple proposition, say $x \in Y$ for *this town has a tavern* and express the quantified proposition by $(x)(x \in Y)$?

The answer is that in this case $x$ would not have any precise significance, even as a place-keeper. If it can stand for just any symbol, i.e., it is a genuine dummy, then the proposition $(x)(x \in Y)$ means *everything is a town,* which though not exactly nonsense is not the sense intended. If on the other hand, $x$ stands for *that town* and nothing else it has ceased to be a place-keeper at all. The overwhelming advantage of the symbolism here adopted is that although the significance of $x$ can range freely under the quantifier (since it is assumed that its place can be taken by any subject at all) yet it has a precise and precisely the same meaning in each part of any proposition which it is used to form. In other words its use enables us to formulate a precise proposition concerning *all towns* without our needing to have a list of all the towns there are in front of us, or even knowing whether such a list exists or could exist.

A further advantage is that this form of hypothesis is susceptible to modification without difficulty. Suppose, for example, we wanted to restrict the hypothesis to English-speaking towns. If the letter $A$ be understood to denote the predicate *English-speaking* the restricted hypothesis would be expressed as:

$$(x)((x \in X \wedge x \in A) \Rightarrow x \in Y)$$

We can of course form the negation of a quantified proposition, and this we denote by $(x)'(\phi x)$; care must be taken to distinguish this from the quantification of a negated matrix, $(x)(\phi' x)$.

The necessary definitions are:

$$(x)'(\phi x) \equiv ((x)(\phi x))' \quad \text{Df} \tag{2.2}$$

$$(x)(\phi' x) \equiv (x)((\phi x)') \quad \text{Df} \tag{2.3}$$

*Example 1*

Render into symbolic form:

(a) That town is not furnished with a tavern.

(b) No town is furnished with a tavern. (Universal denial)

(c) Not every town is furnished with a tavern. (Denial of a universal)

(d) It is not true that all towns other than English-speaking ones are not furnished with taverns. (Denial of a universal denial)

With the notation of the previous paragraphs, the required symbolism is:

(a) $x_0 \notin Y$

(b) $(x)(x \in X \Rightarrow x \notin Y)$

(c) $(x)'(x \in X \Rightarrow x \in Y)$

(d) $(x)'((x \in X \wedge x \notin A) \Rightarrow x \notin Y)$

(c) *The existential quantifier*

It is often convenient to quantify in a different way. Example 1(d) could have been more conveniently expressed by *Some towns, not English-speaking, are furnished with taverns* or, more exactly, *There is at least one*[7] *town which is not English-speaking which is furnished with a tavern.*

Consider an even simpler instance: *There is a tavern in a town.* By this we are led to understand that there exists at least one town, somewhere, which is furnished with a tavern. It is convenient to have a special symbol for such a proposition as this, and with the same meanings for $X$ and $Y$ as before we should write it as:

$$(\exists x)(x \in X \wedge x \in Y) \tag{2.4}$$

which is generally pronounced *There exists an x such that x is an X and x is a Y.*

But the same sense is conveyed by *It is not the case that all towns are unfurnished with taverns* and this we should write in symbols as:

$$(x)'(x \in X \Rightarrow x \notin Y)$$

which in virtue of (1.12) is equivalent to:

$$(x)'(x \in X \wedge x \in Y)'$$

which we take to be the definition of (2.4).

More generally we define the use of the **existential quantifier**, $(\exists x)$, treated again as a single symbol, by:

$$(\exists x)(\phi x) \equiv (x)'(\phi' x) \quad \text{Df} \qquad (2.5)$$

and the negation of this by:

$$(\exists x)'(\phi x) \equiv (x)(\phi' x) \quad \text{Df} \qquad (2.6)$$

Immediate consequences of the definition are:

$$(\exists x)(\phi' x) \equiv (x)'(\phi x) \qquad (2.7)$$

and

$$(\exists x)'(\phi' x) \equiv (x)(\phi x) \qquad (2.8)$$

Note: We have been careful to speak of the use of the symbols $(x)$ and $(\exists x)$ rather than the symbols themselves. This may seem a hair-splitting distinction since in one sense it is only the use of a symbol that is ever capable of definition. What we mean to draw attention to is the fact that these symbols can never stand by themselves on one side of the sign of equivalence ($\equiv$) nor in fact next to any symbol of relationship. This is particularly important in the case of $(\exists x)$. It is often forgotten when questions are debated as to whether such-and-such 'exists' (the Loch Ness Monster, for example) that something must be predicated of such-and-such if the proposition 'such-and-such exists' is to have any meaning.

### (d) *Distribution of quantifiers*

Let $P$ stand for the totality of places where English is spoken, and $Q$ for the totality of places where beer is drunk. Now consider the following proposition:

$$(x)((x \in X \Rightarrow x \in Y) \Rightarrow (x \in P \Rightarrow x \in Q) \qquad (2.9)$$

Since the first subsidiary implication, $x \in X \Rightarrow x \in Y$, has a truth value of 1 whenever $x \notin X$ has, (2.9) asserts of every place which is not a town that if English is spoken there then beer is drunk; and of every place that is a town, provided it has a tavern, then if English is spoken, beer is drunk.

Compare this with the following:

$$(x)(x \in X \Rightarrow x \in Y) \Rightarrow (x)(x \in P \Rightarrow x \in Q) \tag{2.10}$$

This asserts that if every town has a tavern then wherever English is spoken, beer is drunk. A little thought will show that (2.9) and (2.10) are not equivalent. A single instance of a place which is not a town, where English is spoken but beer not drunk will suffice to falsify (2.9), but it will not necessarily falsify (2.10). On the other hand any instance which falsifies (2.10) will necessarily falsify (2.9). In other words, though the propositions are not equivalent, the first implies the second.

This conclusion does not of course depend on the specific meanings given to $X$, $Y$, $P$ and $Q$; nor even on the form of the propositions related by the principal sign of implication in (2.9). The existence of a single $a$ such that $\phi a$ is 'true' and $\psi a$ 'false' falsifies $(x)(\phi x \Rightarrow \psi x)$ without necessarily falsifying $(x)(\phi x) \Rightarrow (x)(\psi x)$, whereas it is necessary for every $\phi a$ to be 'true' and one $\psi a$ 'false' for $(x)(\phi x) \Rightarrow (x)(\psi x)$ to be 'false'; and in the latter case $(x)(\phi x \Rightarrow \psi x)$ is 'false' also.

We cannot therefore assert $(x)(\phi x \Rightarrow \psi x) \equiv ((x)\, \phi x \Rightarrow (x)(\psi x))$; instead we are led to the weaker assertion:

$$(x)(\phi x \Rightarrow \psi x) \Rightarrow ((x)(\phi x) \Rightarrow (x)(\psi x)) \tag{2.11}$$

Combining this with the same proposition but with $\phi$ and $\psi$ interchanged and appealing to (2.13) we reach the conclusion:

$$(x)(\phi x \equiv \psi x) \Rightarrow ((x)(\phi x) \equiv (x)(\psi x)) \tag{2.12}$$

The following results, rigorous proofs of which are deferred until section **2.5**, may be reached by similar reasoning:

$$(x)(\phi x \wedge \psi x) \equiv ((x)(\phi x) \wedge (x)(\psi x)) \tag{2.13}$$

$$(x)(\phi x \wedge p) \equiv ((x)(\phi x) \wedge p) \tag{2.14}$$

where $p$ is a proposition not containing $x$, and

$$(x)(\phi x \vee p) \equiv ((x)(\phi x) \vee p) \tag{2.15}$$

where again $p$ does not contain $x$.

*Example 2*

Infer from the foregoing:

$$(\exists x)(\phi x \vee \psi x) \equiv ((\exists x)(\phi x) \vee (\exists x)(\psi x)) \tag{2.16}$$

The argument is as follows:

$$(x)(\phi x \wedge \psi x) \equiv ((x)(\phi x) \wedge (x)(\psi x)) \tag{2.17}$$

(2.12)

$$(x)'(\phi x \wedge \psi x) \equiv ((x)(\phi x) \wedge (x)(\psi x))' \qquad (2.18)$$

(2.17) and (1.16)

$$(x)'(\phi' x \wedge \psi' x) \equiv ((x)(\phi' x) \wedge (x)(\psi' x))' \qquad (2.19)$$

(2.18: $\phi' x$ for $\phi x$, $\psi' x$ for $\psi x$)

$$(x)'(\phi x \vee \psi x)' \equiv ((x)'(\phi' x) \vee (x)'(\psi' x)) \qquad (2.20)$$

(2.19), (1.11) and (1.12)

$$(\exists x)(\phi x \vee \psi x) \equiv ((\exists x)(\phi x) \vee (\exists x)(\psi x)) \qquad (2.21)$$

(2.20) and (2.5)

*Example 3*

Infer from the foregoing:

$$((x)(\phi x) \vee (x)(\psi x)) \Rightarrow (x)(\phi x \vee \psi x)$$

$$(x)(\phi x \Rightarrow (\phi x \vee \psi x)) \qquad (2.22)$$

(1.7: $\phi x$ for $p$, $\psi x$ for $q$)

$$(x)(\phi x) \Rightarrow (x)(\phi x \vee \psi x) \qquad (2.23)$$

(2.22) and (2.11)

$$(x)(\psi x) \Rightarrow (x)(\phi x \vee \psi x) \qquad (2.24)$$

(2.23: $\phi x$ for $\psi x$, $\psi x$ for $\phi x$)

$$((x)(\phi x) \vee (x)(\psi x)) \Rightarrow (x)(\phi x \vee \psi x) \qquad (2.25)$$

(2.23), (2.24) and (1.27)

*Example 4*

Infer from the foregoing:

$$(\exists x)(\phi x \wedge \psi x) \Rightarrow ((\exists x)(\phi x) \wedge (\exists x)(\psi x))$$

$$((x)(\phi x) \vee (x)(\psi x)) \Rightarrow (x)(\phi x \vee \psi x) \qquad (2.26)$$

(2.25)

$$(x)'(\phi x \vee \psi x) \Rightarrow ((x)(\phi x) \vee (x)(\psi x)' \qquad (2.27)$$

(2.26) and (1.15)

$$(x)'(\phi' x \wedge \psi' x)' \Rightarrow ((x)'(\phi x) \wedge (x)'(\psi x)) \qquad (2.28)$$

(2.27) (1.11) and (1.12)

$$(\exists x)(\phi' x \wedge \psi' x) \Rightarrow ((\exists x)(\phi' x) \wedge (\exists x)(\psi' x)) \qquad (2.29)$$

(2.28) (2.5) and (2.6)

$$(\exists x)(\phi x \wedge \psi x) \Rightarrow ((\exists x)(\phi x) \wedge (\exists x)(\psi x)) \qquad (2.30)$$

(2.29: $\phi' x$ for $\phi x$, $\psi' x$ for $\psi x$)

Notice that in the last two examples the distribution of the quantifier is in opposite senses. The single quantifier appears on the right-hand side of the sign of implication in the first case and on the left-hand side in the second. This is characteristic of the duality which obtains between the universal and existential quantifiers.

*Example 5*

Infer from the foregoing:

$$((x)(\phi x) \wedge (\exists x)(\psi x)) \Rightarrow (\exists x)(\phi x \wedge \psi x)$$

$$(((x)(\phi x) \wedge (\exists x)(\psi x))' \vee (\exists x)(\phi x \wedge \psi x))$$

$$\equiv ((x)'(\phi x) \vee (\exists x)'(\psi x)) \vee (\exists x)(\phi x \wedge \psi x) \tag{2.31}$$

(1.11: $(x)(\phi x)$ for $p$, $(\exists x)(\psi x)$ for $q$)

$$\equiv ((x)'(\phi x) \vee (\exists x)(\phi x \wedge \psi x)) \vee (\exists x)'(\psi x) \tag{2.32}$$

(1.18: $(\exists x)(\phi x \wedge \psi x)$ for $r$)

$$\equiv ((\exists x)(\phi' x) \vee (\exists x)(\phi x \wedge \psi x) \vee (\exists x)'(\psi x) \tag{2.33}$$

(2.6)

$$\equiv (\exists x)(\phi' x \vee (\phi x \wedge \psi x)) \vee (\exists x)'(\psi x) \tag{2.34}$$

(2.21)

$$\equiv (\exists x)(\phi' x \vee \psi x) \vee (\exists x)'(\psi x) \tag{2.35}$$

(1.47: $\phi x$ for $p$, $\psi x$ for $q$)

$$\equiv (\exists x)(\phi' x) \vee (\exists x)(\psi x) \vee (\exists x)'(\psi x) \tag{2.36}$$

(2.21).

But the last proposition in this sequence of equivalences[7] is tautologous, so the first is; and the first is just the inference it is required to make, namely

$$((x)(\phi x) \wedge (\exists x)(\psi x)) \Rightarrow (\exists x)(\phi x \wedge \psi x) \tag{2.37}$$

*Example 6*

Prove that $\phi a \Rightarrow (\exists x)(\phi x)$, where $a$ is some possible signification of $x$.

$$(x)(\phi x) \Rightarrow \phi a \tag{2.38}$$

Premiss – see Note 14

$$\phi' a \Rightarrow (x)'(\phi x) \tag{2.39}$$

(1.15) and (2.38)

$$\phi a \Rightarrow (x)'(\phi' x) \tag{2.40}$$

(2.39: $\phi'$ for $\phi$)

$$\phi a \Rightarrow (\exists x)(\phi x) \tag{2.41}$$

Definition

*Examples 7 and 8*

Use (2.14) and (2.15) to prove

$$p \wedge (\exists x)(\phi x) \equiv (\exists x)(p \wedge \phi x) \tag{2.42}$$

$$p \vee (\exists x)(\phi x) \equiv (\exists x)(p \vee \phi x) \tag{2.43}$$

where $p$ is a proposition not containing $x$. These are left as exercises.

---

**Exercises**

2.1 Work Examples 7 and 8.

2.2 Use the notation suggested to put the following propositions in symbolic form:

(a) Some English towns are furnished with taverns.
(b) No towns are both English and French.
(c) Some French towns are English-speaking.
$V \equiv$ French; $W \equiv$ English; $X \equiv$ Towns; $Y \equiv$ furnished with a tavern;
$Z \equiv$ English-speaking.

2.3 Prove $((x)(\phi x \equiv \psi x)) \not\equiv ((x)(\phi x) \equiv (x)(\psi x))$

2.4 Prove $((\exists x)(\phi x \Rightarrow \psi x)) \equiv ((x)(\phi x) \Rightarrow (\exists y(\psi y))$ and distinguish this from $(\exists x)(\phi x \Rightarrow \psi x) \equiv (x)(\phi x \Rightarrow (\exists y)(\psi y))$. Is the latter tautologous?

2.5 Extend the results of (2.13), (2.21), (2.25) and (2.30) to three or more propositional matrices.

2.6 Prove $((\exists x)(\phi x) \wedge (\exists x)(\psi x)) \Rightarrow (\exists x)(\phi x \vee \psi x)$.

2.7 Prove $(x)(\phi x \Rightarrow \psi x) \Rightarrow ((\exists x)(\phi x) \Rightarrow (\exists x)(\psi x))$.

---

## 2.3 The calculus of sets

It is sometimes said that the idea of a set is 'intuitive'. But is it? We cannot arrive at a set simply by aggregating its members. The aggregate of all English parishes is the same thing as the aggregate of all English counties – namely England. But the two sets are not the same. Unless we are content therefore with the definition of a set as a mere list of its members – and plainly this is one possible definition – we are thrown back on the notion of a typical member. And in fact definition in terms of such a member is what the ordinary use of language suggests. To 'belong to the set of cats' means simply to be a cat, i.e., to have at least all the qualities which serve to define a cat, the appropriate feline genes or whatever they may be.

Suppose the letter $T$ to stand for that subject – real or imaginary – which enjoys all the predicates of a town and no others. $T$, in other words, is to stand for a typical town.[8] Then to belong to the set, $X$, of towns means to enjoy at least all the predicates of $T$. Accordingly we define membership in $X$ as follows:

$$x \in X \equiv (i)(T \in i \Rightarrow x \in i) \quad \text{Df} \tag{2.44}$$

The importance of this definition is that it makes possible the transition from the notion of a subject $T$, to that of a **class** or **set**, $x$. It also makes it possible to link up what has been said in this chapter about quantified propositions with what was said in the last chapter about propositions in general, and thus to discover how parts of propositions, and more especially their predicates, may be manipulated within the general framework laid down in chapter 1.

(a) *Relation of sets to propositions*

It will be shown in what follows that sets, or more precisely the symbols denoting them, can in fact be manipulated according to rules almost exactly analogous to those which govern the manipulation of propositions. This is not altogether surprising since membership in a set is just the form which every simple proposition is now supposed to take.

(b) *Complementary sets*

Consider first the negation of the proposition $x \in i$, what we have agreed to denote by $x \notin i$. It is clear that this can be used to define a set complementary to $X$ as defined in (2.44). This complementary set we denote by $X'$. A subject will be a member of this set if and only if it lacks some predicate of $T$. Thus:

$$x \in X' \equiv (\exists i)(T \in i \wedge x \notin i) \quad \text{Df} \tag{2.45}$$

also

$$x \in X \equiv (\exists i)'(T \in i \Rightarrow x \in i)' \tag{2.46}$$

(2.44) and (2.8)

$$x \in X \equiv (\exists i)'(x \notin i \wedge T \in i) \tag{2.47}$$

(2.46) and (1.12)

so

$$x \in X' \equiv (x \in X)'$$

(2.47) and (2.45)

The foregoing argument is in no way dependent on the choice of $x$. We may therefore conclude:

$$(x)(x \in X' \equiv x \notin X) \tag{2.48}$$

and indeed we might well have made this the basis of our definition of membership in $X'$, the **complement** of $X$, rather than (2.45). The analogy with the negation of a proposition is obvious.

(c) *Union and intersection of sets*

As in the definition of the complement of a set we made use of the notion of negation, so in the definition of the **union** of two sets, $X$, $Y$ we make use of the notion of alternation:

$$x \in X \cup Y \equiv (x \in X \vee x \in Y) \quad \text{Df} \tag{2.49}$$

The corresponding definition of membership in the **intersection** of two sets, $X$, $Y$ is similarly,

$$x \in X \cap Y \equiv (x \in X \wedge x \in Y) \quad \text{Df} \tag{2.50}$$

It should be noted that all these definitions – (2.45), (2.49) and (2.50) – are definitions in use, defining membership in certain sets rather than the sets themselves. The symbols $\cup$ and $\cap$ conveniently similar to $\vee$ and $\wedge$ are sometimes pronounced 'cup' and 'cap'.

(d) *The tautological forms*

On the basis of the foregoing definitions it is possible to establish set-membership tautologies. These are exactly analogous to the tautologies of section **1.3**(d) and may be derived from them. From (1.12) we can infer:

$$x \in X \cap Y \equiv x \in (X' \cap Y')'$$

as follows:

$$(x \in X \vee x \in Y)' \equiv (x \in X' \wedge x \in Y') \qquad (2.52)$$

(1.12: $x \in X$ for $p$, $x \in Y$ for $q$)

$$(x \in X \cup Y)' \equiv x \in X' \cap Y' \qquad (2.53)$$

(2.52), (2.49) and (2.50) Definition

$$x \in (X \cup Y)' \equiv x \in X' \cap Y' \qquad (2.54)$$

(2.53) and (2.48) Definition

Using (1.11) in place of (1.12) a similar argument leads to the conclusion

$$x \in (X \cap Y)' \equiv x \in X' \cup Y' \qquad (2.55)$$

We can in this way reach a number of tautologies analogous to those listed in section **1.3**(d). Instead of writing them out in full, however, it is convenient to put them in short-hand form. The two tautologies above would then be written as

$$(X \cup Y)' \equiv X' \cap Y' \qquad (2.56)$$

and

$$(X \cap Y)' \equiv X' \cup Y' \qquad (2.57)$$

These 'short-hand equivalences' are a way of expressing the equivalences of pairs of propositions written in the subject–copula–predicate form. Although they can be freely used as *rules* for the manipulation of letters denoting 'sets' it is important to remember that in fact such letters have no significance in isolation. When we manipulate them what we are really doing is to manipulate propositions, $x \in X$, $x \subset Y$ and so on, which they are a part.

All these tautologies which we assume listed here as in section **1.3**(d) can be illustrated by means of Venn diagrams.[9] In such diagrams membership of an element in a set or the joining of a subject to a predicate is represented by the location of a point within a circle or other closed curve; membership in the complementary set being represented by location within the remaining area of the diagram. A reader who is not familiar with the use of such diagrams could not do better than pause at this point and select a dozen or more of these tautologies and sketch diagrams to illustrate them.

Figures 2.1 and 2.2 are the illustrations appropriate to (2.54) and (2.55) in the cases where the circles whose interiors represent $X$ and $Y$ overlap. Some thought should be given to the cases in which they do not.

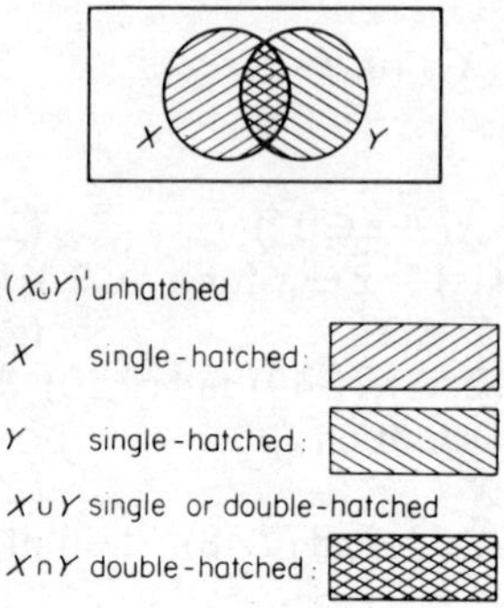


Fig. 2.1

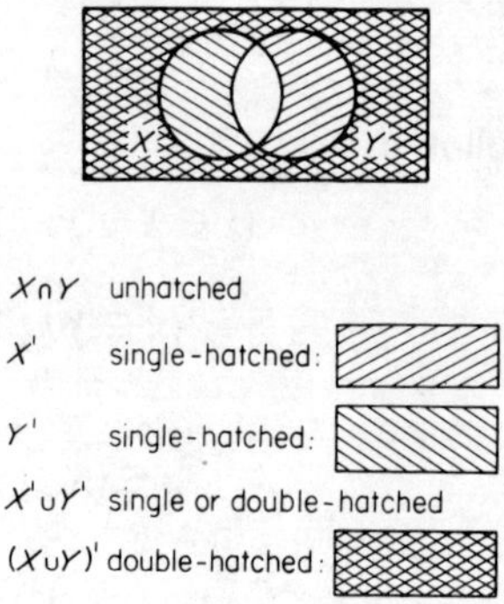


Fig. 2.2

### (e) *Set inclusion and set equivalence*

It has been pointed out that tautologies similar to those in section **1**.3(d) can be obtained from them by substituting $x \in X$ for $p$, etc.; in many cases the resulting tautology could have been derived in its short-hand form simply by substituting the letters $X, Y, Z$ for $p, q, r$; $\cup$ for $\vee$ and $\cap$ for $\wedge$. Are there any exceptions to this rule? The answer is that there are, and they arise in virtue of the fact that tautologies involving sets contain a concealed variable $x$, and a concealed quantifier governing that variable.

The exceptions arise, in particular, when the propositional tautology which we take as our point of departure involves the sign of implication; the corresponding connective between sets expresses inclusion, a relation which obtains when every member of one set is also a member of the other. The sign denoting inclusion[10] is $\subset$, and the definition of its use is as follows:

$$X \subset Y \equiv (x)(x \in X \Rightarrow x \in Y) \quad \text{Df} \qquad (2.58)$$

It is noticeable that at this point the exact analogy between the rules for manipulating propositions and those for manipulating sets breaks down. We do not define $X \subset Y$ as $X' \cup Y$ – though we could have done so had we wished; that would have made $X \subset Y$ stand for a predicate. Instead we make it stand for a proposition whose assertion is equivalent to the assertion that of every subject $x$ either $x \notin X$ or $x \in Y$. In the language of the Venn diagram, this is equivalent to the assertion that every point of the diagram is either outside the circle denoted by $X$ or inside the circle denoted by $Y$ or both. In other words the interior of $X$ lies wholly within (or on) the interior of $Y$ (Fig. 2.3).

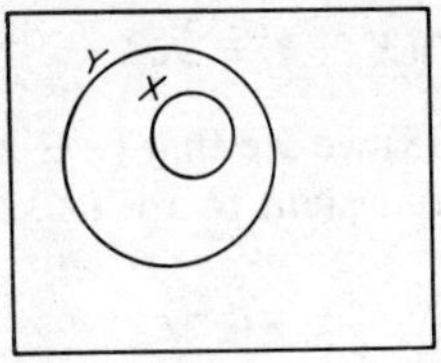


$X \subset Y$

Fig. 2.3

It follows that when a propositional tautology contains an implication within an implication, the corresponding tautology in the calculus of sets will contain not an inclusion within an inclusion but an inclusion within an implication. (1.25) for example will lead to:

$$X \subset Y \Rightarrow X \cap Z \in Y \cap Z \tag{2.59}$$

It remains to show that if $X$, $Y$ are sets such that each includes the other then they are equivalent in the sense that membership in the one is equivalent to membership in the other. We first require a definition:

$$X \equiv Y \equiv (x)(x \in X \equiv x \in Y) \quad \text{Df} \tag{2.60}$$

The proof follows:

$$(x)(\phi x \wedge \psi x) \equiv (x)(\phi x) \wedge (x)(\psi x) \tag{2.61}$$

(2.13)

$$(x)((x \in X \Rightarrow x \in Y) \wedge (x \in Y \Rightarrow x \in X)) \equiv ((x)(x \in X \Rightarrow x \in Y) \wedge (x)(x \in Y \Rightarrow x \in X)) \tag{2.62}$$

(2.61: $x \in X \Rightarrow x \in Y$ for $\phi x$, $x \in Y \Rightarrow x \in X$ for $\psi x$)

$$(x)(x \in X \equiv x \in Y) \equiv (X \subset Y \wedge Y \subset X) \tag{2.63}$$

(2.62), (1.143): $x \in X$ for $p$, $x \in Y$ for $q$) and (2.58: Definition)

$$(X \equiv Y) \equiv (X \subset Y \wedge Y \subset X) \tag{2.64}$$

(2.63: Definition)

(f) *Universal and null sets*

From (1.36) we can derive the tautology:

$$X \cup X' \equiv Y \cup Y' \tag{2.65}$$

Let us suppose $Y$ to be an arbitrarily chosen but fixed set and denote $Y \cup Y'$ by $\vee$. Since $X$ is also chosen arbitrarily we can write:

$$(X)(X \cup X' \equiv \vee) \tag{2.66}$$

and similarly, denoting $Y \cap Y'$ by $\wedge$

$$(X)(X \cap X' \equiv \wedge) \tag{2.67}$$

Taking $x \in X$ for $p$ in (1.35) we see that $(x \in X) \vee (x \in X)'$ is a tautology; and as this is equivalent to $x \in (X \cup X')$ we may add to our list of tautologies:

$$x \in \vee \tag{2.68}$$

and similarly:

$$x \notin \wedge \tag{2.69}$$

$\vee$, $\wedge$ are said to denote the **universal** and **null** sets, respectively. Membership in $\vee$ is that which can be predicated of every subject; membership in $\wedge$ is that which can be predicated of no subject.

(g) *Subsets*

An important theorem is: $Y \subset Z \Rightarrow (X \subset Y \Rightarrow X \subset Z)$ (2.70)

This states that if one class, $Y$, is included in another, $Z$, then the inclusion of any set in $Y$ implies its inclusion in $Z$.

The proof is as follows:

$$(x)((x \in Y \Rightarrow x \in Z) \Rightarrow ((x \in X \Rightarrow x \in Y) \Rightarrow (x \in X \Rightarrow x \in Z))) \tag{2.71}$$

(1.26: $x \in Y$ for $p$, $x \in Z$ for $q$, $x \notin X$ for $r$)

$$(x)(x \in Y \Rightarrow x \in Z) \Rightarrow (x)((x \in X \Rightarrow x \in Y) \Rightarrow (x \in X \Rightarrow x \in Z)) \tag{2.72}$$

(2.71) and (2.11: $x \in Y \Rightarrow x \in Z$ for $\phi x$ $(x \in X \Rightarrow x \in Y) \Rightarrow (x \in X \Rightarrow x \in Z)$ for $\psi x$)

$$(x)((x \in X \Rightarrow x \in Y) \Rightarrow (x \in X \Rightarrow x \in Z)) \Rightarrow ((x)(x \in X \Rightarrow x \in Y) \Rightarrow (x)(x \in X \Rightarrow x \in Z)) \tag{2.73}$$

(2.11: $x \in X \Rightarrow x \in Y$ for $\phi x$, $x \in X \Rightarrow x \in Z$ for $\psi x$)

$$(x)(x \in Y \Rightarrow x \in Z) \Rightarrow ((x)(x \in X \Rightarrow x \in Y) \Rightarrow (x)(x \in X \Rightarrow x \in Z)) \tag{2.74}$$

(2.72) and (2.73)

$$Y \subset Z \Rightarrow (X \subset Y \Rightarrow X \subset Z) \tag{2.70}$$

(2.74) and (2.58: Definition)

A similar argument proves:

$$Y \in Z \Rightarrow (Z \subset X \Rightarrow Y \subset X) \qquad (2.75)$$

Immediate inferences[11] from (2.70) and (2.75) are:

$$Y \subset Z \Rightarrow (X)(X \subset Y \Rightarrow X \subset Z) \qquad (2.76)$$
$$Y \subset Z \Rightarrow (X)(Z \subset X \Rightarrow Y \subset X) \qquad (2.77)$$

The converse[12] of these theorems follow readily, as follows:

$$(X)(X \subset Y \Rightarrow X \subset Z) \Rightarrow (Y \subset Y \Rightarrow Y \subset Z) \qquad (2.78)$$

(Premiss:[13] $Y$ for $a$)

$$Y \subset Y \qquad (2.79)$$

(1.3: $x \in Y$ for $p$) and (2.58)

$$(X)(X \subset Y \Rightarrow X \subset Z) \Rightarrow Y \subset Z \qquad (2.80)$$

(2.78) and (2.79: RI)

$$(X)(Z \subset X \Rightarrow Y \subset X) \Rightarrow (Z \subset Z \Rightarrow Y \subset Z) \qquad (2.81)$$

(Premiss: $Z$ for $a$)

$$Z \subset Z \qquad (2.82)$$

(1.3: $x \in Z$ for $p$) and (2.58)

$$(X)(Z \subset X \Rightarrow Y \subset X) \Rightarrow Y \subset Z \qquad (2.83)$$

(2.81) and (2.82: RI)

Taking (2.76) and (2.77) each with its converse, we are led to:

$$Y \subset Z \equiv (X)(X \subset Y \Rightarrow X \subset Z) \qquad (2.84)$$

and

$$Y \subset Z \equiv (X)(Z \subset X \Rightarrow Y \subset X) \qquad (2.85)$$

If we now interchange $Y$ and $Z$ in (2.84) and combine the result with (2.84) itself, we are led to the further conclusion:

$$(Y \equiv Z) \equiv (X)(X \subset Y \equiv X \subset Z) \qquad (2.86)$$

and, treating (2.85) similarly:

$$(Y \equiv Z) \equiv (X)(Y \subset X \equiv Z \subset X) \qquad (2.87)$$

These results should be carefully distinguished from propositions which are similar in form. (2.84) looks like the definition of inclusion, (2.58: $Y$ for $X$, $Z$ for $Y$), i.e.:

$$Y \subset Z \equiv (x)(x \in Y \Rightarrow x \in Z) \qquad (2.88)$$

The difference does not lie in the substitution of $x$ for $X$, for the variable of quantification is only a dummy and any symbol will serve the purpose. The difference lies in the substitution of $\in$ for $\subset$. The use of the latter assures us that $X$ stands not for any subject but for a set.

Whether sets can themselves be subjects and therefore members of sets in their turn is a question to which we must give further consideration (section **2.5**). But if we assume for the moment that sets can be members of sets it is clear that membership in a set and inclusion in it are quite distinct relationships.

Suppose for the moment that the set whose members are $x, y, z$ is denoted by $\{x, y, z\}$ and similarly any other set whose members can be listed. It is then apparent that $\{b, c\}$ is included in $\{a, b, c\}$ but is not a member of it; it is a member of $\{\{b, c\}, \{c, a\}, \{a, b\}\}$, but not included in it; and it is both a member of and included in $\{\{b, c\}, b, c\}$.

If $Y \subset Z$, $Y$ is referred to as a **subset** of $Z$, and if we can assert further, $(Y \equiv \wedge)'$ and $(Z \equiv Y)'$, $Y$ is termed a **proper subset**. The foregoing theorems tell us that in either case every subset of $Y$ is a subset of $Z$ and that of every set of which $Z$ is a subset $Y$ is also a subset.

(h) *Unit sets*

Although we shall not have occasion to use it just yet, it is convenient at this point to give the definition of membership in a **unit set**, that is the set whose sole member is a given subject, $a$.

$$x \in [a] \equiv (i)(a \in i \equiv x \in i) \quad \text{Df} \tag{2.89}$$

The similarity with (2.44), the definition of membership in general, should be noted.

*Example 9*

Derive, in the calculus of sets, a tautology to correspond to (1.13), and illustrate by means of a Venn diagram.

(1.13) is:

$$(p \equiv q') \equiv (p \equiv q)'$$

thus

$$(x \in X \equiv x \notin Y) \equiv (x \in X \equiv x \in Y)' \tag{2.90}$$

(1.13: $x \in X$ for $p$,
$x \in Y$ for $q$)

$$(x \in X \equiv x \in Y)' \equiv ((x \in X \wedge x \notin Y) \vee (x \notin X \wedge x \in Y)) \tag{2.91}$$

(1.48: $x \in X$ for $p$,
$x \in Y$ for $q$)

$$((x \in X \wedge x \notin Y) \vee (x \notin X \wedge x \in Y)) \equiv (x \in \vee \equiv (x \in X \cap x \notin Y) \vee (x \in X \wedge x \in Y)) \tag{2.92}$$

$$(1.49\colon (x \in X \wedge x \notin Y) \vee (x \notin X \wedge x \in Y) \text{ for } q)$$

$$(x \in X \equiv x \notin Y) \equiv (x \in \mathcal{V} \equiv (x \in X \wedge x \in Y) \vee (x \notin X \wedge x \in Y)) \tag{2.93}$$

(2.90), (2.91) and (2.92)

$$(X \equiv Y') \equiv (\mathcal{V} \equiv (X \cap Y') \cup (X' \cap Y)) \tag{2.94}$$

(2.93), (2.13) and (2.60: Definition)

The Venn diagram is shown in Figure 2.4.

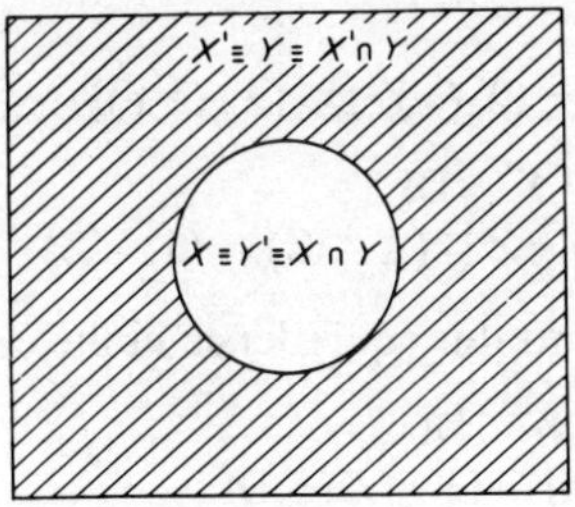


Fig. 2.4

*Example 10*

Show that the inclusion of $X$ in $Y$ may be expressed as either (a) the equivalence of $X$ and $X \cap Y$ or (b) the equivalence of $Y$ and $X \cup Y$.

(a)
$$(p \Rightarrow q) \equiv (p \equiv (p \wedge q)) \tag{2.95}$$

(1.37)

$$(x \in X \Rightarrow x \in Y) \equiv (x \in X \equiv (x \in X \wedge x \in Y)) \tag{2.96}$$

(2.95: $x \in X$ for $p$, $x \in Y$ for $q$)

$$X \subset Y \equiv (X \equiv X \cap Y) \tag{2.97}$$

(2.96), (2.13) and (2.60: Definition)

(b)
$$(p \Rightarrow q) \equiv (q' \Rightarrow p') \tag{2.98}$$

(1.15)

$$(q' \Rightarrow p') \equiv (q' \equiv (p' \wedge q')) \tag{2.99}$$

(1.37: $q'$ for $p$, $p'$ for $q$)

$$(q' \equiv (p' \wedge q')) \equiv (q \equiv (p \vee q)) \tag{2.100}$$

(1.16: $p \vee q$ for $p$)

$$(p \Rightarrow q) \equiv (q \equiv (p \vee q)) \tag{2.101}$$

(2.98), (2.99) and (2.100: RI)

$$(x \in X \Rightarrow x \in Y) \equiv (x \in Y \equiv (x \in X \vee x \in Y)) \tag{2.102}$$

(2.101: $x \in X$ for $p$, $x \in Y$ for $q$)

$$X \subset Y \equiv (Y \equiv X \cup Y) \tag{2.103}$$

(2.102), (2.13) and (2.60: Definition)

## Exercises

2.8 Make a selection of the tautologies in section **1.3**(d); derive corresponding propositions in the calculus of sets, and illustrate where possible with Venn diagrams.

2.9 Obtain expressions in the calculus of sets to assert that (a) $X$ has some elements in common with $Y$; and (b) $X$ has no elements in common with $Y$. Are your expressions symmetrical, i.e., are they unaffected by the interchange of $X$ and $Y$?

2.10 Prove:
$$(X \equiv Y) \equiv (\Lambda \equiv (X \cap Y') \cup (X' \cap Y))$$

2.11 If $X - Y \equiv X \cap Y'$ Df,
prove: $(X - Y) - Z \equiv (X - Z) - (Y - Z)$

What corresponds to this equivalence in the calculus of propositions?

2.12 If $X \equiv Y - X \cap Y'$ Df
and $X + Y \equiv (X - Y) \cap (Y - X)$ Df
prove:
(a) $X + Y \equiv Y + X$
(b) $X + (Y + Z) \equiv (X + Y) + Z$
(c) $X + X \equiv \Lambda$
(d) $X + \Lambda \equiv X$
(e) $(X \equiv Y) \equiv (X + Y \equiv \Lambda)$

Verify that these five equivalences are consistent with the following interpretation:

$X \equiv$ last digit of binary $x$
$Y \equiv$ last digit of binary $y$
$X + Y \equiv$ last digit of binary $(x + y)$
$\Lambda \equiv 0$ (i.e., zero)

2.13 Simplify $(X \cup (Y \cap Z')) \cap (Y \cup (Z \cap X'))$ and hence prove:
$$(X \cup (Y \cap Z')) \cap (Y \cup (Z \cap X')) \cap (Z \cup (X \cap Y')) \equiv X \cap Y \cap Z$$

2.14 Prove: $((X \subset Z) \wedge (Y \Lambda Z)) \equiv X \cup Y \subset Z$

## 2.4 Argument in the calculus of sets

Every valid inference depends upon a tautologous implication (section **1.4**(a). This is as true in the calculus of sets as in the calculus of propositions and for the same reasons.

(a) *Review of arguments*

All the arguments considered in the previous chapter will have their analogies in the calculus of sets. In some cases it will be easiest to obtain the necessary implication from conclusions already reached in this chapter; in others to go back to first principles. With this in mind let us review the main arguments so far considered (see section **1.4**).

(i) The rule of detachment

This depends on $((p \Rightarrow q) \wedge p) \Rightarrow q$

The corresponding inference in the calculus of sets is:

$$\begin{array}{c} X \subset Y \\ x \in X \\ \hline x \in Y \end{array}$$

The argument to justify this runs as follows:

$$X \subset Y \tag{2.104}$$

Premiss

$$X \subset Y \equiv (y)(y \in X \Rightarrow y \in Y) \tag{2.105}$$

(2.58: $y$ for $x$: Definition)

$$(y)(y \in X \Rightarrow y \in Y) \tag{2.106}$$

(2.104) and (2.105)

$$x \in X \Rightarrow x \in Y \tag{2.107}$$

(2.106) RI[11]

$$x \in X \tag{2.108}$$

Premiss

$$x \in Y \tag{2.109}$$

(2.107) and (2.108)

(ii) The syllogism

This depends on $((p \Rightarrow q) \wedge (q \Rightarrow r)) \Rightarrow (p \Rightarrow r)$

The corresponding inference is:

$$\begin{array}{c} X \subset Y \\ Y \subset Z \\ \hline X \subset Z \end{array}$$

The argument in justification is:

$$Y \subset Z \tag{2.110}$$

Premiss

$$Y \subset Z \equiv (W)(W \subset Y \Rightarrow W \subset Z) \tag{2.111}$$

(2.70: $W$ for $X$)

$$(W)(W \subset Y \Rightarrow W \subset Z) \tag{2.112}$$

(2.110) and (2.111)

$$X \subset Y \Rightarrow X \subset Z \tag{2.113}$$

(2.112) RI[11]

$$X \subset Y \tag{2.114}$$

Premiss

$$X \subset Z \tag{2.115}$$

(2.113) and (2.114)

(iii) The sorites

This depends on $((p \Rightarrow q) \wedge (q \Rightarrow r) \wedge (r \Rightarrow s)) \Rightarrow (p \Rightarrow s)$

The corresponding inference is

$$\begin{array}{c} X \subset Y \\ Y \subset Z \\ Z \subset W \\ \hline X \subset W \end{array}$$

The argument amplifies that of (ii) just as that of section **1.4**(a)(iii) amplifies that of section **1.4**(a)(ii).

*Example 11*

Illustrate the arguments of section **2.4**(a).

So far as sections **2.4**(a)(i) and (ii) are concerned examples already cited in sections **1.4**(a)(i) and (ii) will, with some modification, serve the purpose.

Take section **2.4**(a)(ii) first. Let $X$ stand for the class of grandfathers; $Y$ for the class of fathers, and $Z$ for the class of adults.

The premisses are:

*All grandfathers are fathers*
*All fathers are adults.*

The conclusion is:

*All grandfathers are adults.*

In illustration of section **2.4**(a)(i) let $X$ stand for the class of things not mammals; $Y$ for the class of things that lay eggs; and let $x$ stand for a bird.

It then appears that the premisses are:

*The class of things not mammals is included in the class of things that lay eggs.*
*A bird is not a mammal.*

And the conclusion is:

*A bird lays eggs.*

If, however, we were to substitute *corkscrew* for *bird*, our second premiss would be equally true, for it is a fact that a corkscrew is not a mammal. And we should be led to the conclusion that a corkscrew lays eggs!

This furnishes a good illustration of the dangers involved in making the transition from simple propositions to those expressed in the subject–copula–predicate form. For in fact of course our first premiss is false unless we restrict *things* to *females of the animal kingdom* which is what the argument is really about.

This we can do quite easily by taking $Z$ to stand for this class, the so-called **concourse** of the argument, and conjoining the proposition $Y \subset Z$ to both premiss and conclusion. The inference then is:

$$\frac{\begin{array}{c} Y \subset Z \wedge X \subset Y \\ x \in X \end{array}}{Y \subset Z \wedge x \in Y}$$

As an illustration of the sorites we may take a celebrated example from Lewis Carroll:[14]

Let $X$ stand for the class of babies.
Let $Y$ stand for the class of illogical persons.
Let $Z$ stand for the class of despised persons.
Let $W$ stand for the class of persons unable to manage a crocodile.[15]

The premisses are:

*Babies are illogical persons*
*Illogical persons are despised*
*Despised persons are unable to manage a crocodile.*[15]

And the conclusion is:

*Babies are unable to manage a crocodile.*

## (b) *Symbolic form*

We have now reached the point at which the whole of traditional logic, that is the logic of Aristotle, may be expressed in symbolic form. Appendix I (p. 127) is devoted to showing how this may be done in summary form.

## (c) *The construction of a class of given membership*

We note here that just as we were able to render any complex proposition into its complete disjunctive and conjunctive normal forms by using the rules summarized in (1.19)–(1.22) and (1.31)–(1.34), so now we can render any complex class into similar forms; and, conversely, construct a class whose members satisfy any conditions imposed.

*Example 12*

A woman may draw a pension if either (a) she is over 65 or (b) she is over 60 and a widow or (c) she is over 55, a widow and not working. Express in full disjunctive normal form the class of women who may draw pensions.

Let $A$ stand for the class of women over 65.
Let $B$ stand for the class of women over 60.
Let $C$ stand for the class of women over 55.

so that $A \subset B$ (or $A \equiv A \cap B$) and $B \subset C$ (or $B \equiv B \cap C$), and therefore $A \equiv A \cap B \cap C$.

Let $P$ stand for the class of widows and $Q$ for the class of women not working. Let $X$ stand for the required class. Then:

$$
\begin{aligned}
X &\equiv A \cup (B \cap P) \cup (C \cap P \cap Q) \\
&\equiv (A \cap B \cap C) \cup (B \cap C \cap P) \cup (C \cap P \cap Q) \\
&\equiv ((A \cap B \cap C) \cup (P \cap P') \cup (Q \cap Q')) \cup ((B \cap C \cap P) \cup \\
&\quad (A \cap A') \cup (Q \cap Q')) \cap ((C \cap P \cap Q) \cup (A \cap A') \cup (B \cap B'))
\end{aligned}
$$

Now $A \cap B' = \Lambda$, so we can ignore any intersection containing both $A$ and $B'$. So, using the distributive law repeatedly and ignoring repeated terms, we reach the conclusion:

$$
\begin{aligned}
X \equiv{} & (A \cap B \cap C \cap P \cap Q) \cup (A \cap B \cap C \cap P \cap Q') \cup (A \cap B \cap C \cap \\
& P' \cap Q) \cup (A \cap B \cap C \cap P' \cap Q') \cup (A' \cap B \cap C \cap P \cap Q) \cup \\
& (A' \cap B \cap C \cap P \cap Q') \cup (A' \cap B' \cap C \cap P \cap Q)
\end{aligned}
$$

which is the full disjunctive normal form.

*Example 13*

If a woman may draw a pension if and only if she is excluded from just one of the five classes denoted by $A, B, C, P, Q$ in the previous example, express in its simplest form the class of women who can draw pensions.

Since $A \cap B' \equiv B \cap C' \equiv \Lambda$ we can again ignore any intersection containing these terms. Thus if $Y$ denotes the required class:

$$
\begin{aligned}
Y &\equiv (A \cap B \cap C \cap P \cap Q') \cup (A \cap B \cap C \cap P' \cap Q) \cup (A' \cap B \cap C \cap \\
&\quad P \cap Q) \\
&\equiv (A \cap P \cap Q') \cup (A \cap P' \cap Q) \cup (A' \cap B \cap P \cap Q)
\end{aligned}
$$

i.e., a woman may draw a pension if she is over 65 and either not widowed or working (but not both), or if she is between 60 and 65, widowed and not working.

**Exercises**

2.15 Construct an argument based on the following premisses and draw as many conclusions as you can; note that no restriction is placed upon the number of tutors (*A*, *B*, *C*. . .) by which a student may be taught.

(a) Either a student is not taught by either *A* or *C* or he fails to pass.

(b) Unless a student is not taught by *B* he fails to pass.

(c) Only if a student is not taught by both *C* and *D* does he pass.

How would you introduce into the argument the premiss that there are only four tutors *A*, *B*, *C* and *D*; and how would this premiss affect your conclusions? (Langer: adapted)

2.16 Premiss: *It is always right to do what one believes to be right.*
Premiss: *What one believes to be right is sometimes wrong.*
Conclusion: *It is sometimes right to do what is wrong.*

Is this a valid syllogism? If so, express it in appropriate symbols.

2.17 Examine the formal validity of the following argument:
The proof that political success is not dependent on a public school education is that there are many successful politicians who expressly set out to abolish the public-school system; and no-one would champion such a cause if he had himself been educated at a public-school. (London)

2.18 Repeat question 2.7 with the following:
Obviously one should only punish those who are guilty of a crime; and this shows that it is not true that all criminals are psychologically abnormal, for the abnormal should never be punished. (London)

2.19 Test the validity of the following arguments, having first expressed both premisses and conclusion in symbolic form.

(a) Premiss: *All witnesses are prejudiced.*
Premiss: *Some witnesses are not liars.*
Conclusion: *Some liars are not prejudiced.*

(b) Premiss: *All witnesses are prejudiced.*
Premiss: *Some liars are not prejudiced.*
Conclusion: *Some liars are not witnesses.*

(c) Premiss: *All liars are prejudiced.*
Premiss: *Some witnesses are not liars.*
Conclusion: *Some witnesses are not prejudiced.*

Is the validity of the argument affected in any case if the class of liars is the null class? (Suppes: adapted)

2.20 Examine the validity of the following:
That it is not true that the motorist is always to blame for fatal road accidents is shown by the fact that all fatal accidents on the road are avoidable if only appropriate measures are taken. And many accidents due to faulty road surfaces could be avoided if appropriate measures were taken. And if an accident – fatal or not – is due to a poor road surface, the motorist can hardly be held to blame. (London)

## 2.5 Axioms of quantification and the question of identity

(This section may be omitted at first reading, without loss of continuity.)

Two important questions were deliberately side-stepped in sections (2.1–2.4). The first is whether every 'atomic' proposition – every proposition, that is to say, which cannot be expressed as the conjunction and/or negation of other propositions – has the same form, namely $x \in y$; and, if so, for what precisely do the symbols $x$, $y$ stand? Can both these symbols, for example, denote sets, it being borne in mind that a set is itself only a logical fiction, something to be defined in use?

The second question is whether, if we adopt the axiomatic approach recommended in section (1.6), any further axiom or axioms will be needed once it be granted that propositions (although 'atomic') may have a form which makes it possible to consider them part by part.

### (a) *Axiom of quantification*

Let us take the second question first. We propose to take as axiom of quantification (2.116) below and show that all the theorems so far made use of may be obtained with its help:

$$(x)(\phi x) \Rightarrow \phi a^{16} \tag{2.116}$$

Writing $\phi'$ for $\phi$ and using the law of negative implication (section **1.6**(d) an immediate inference is:

$$\phi a \Rightarrow (\exists x)(\phi x) \tag{2.117}$$

Here, as in (2.116), $\phi a$ stands for that proposition which is equivalent to $\phi x$ when $x$ is given the significance of $a$.

The axiom means that $(x)(\phi x)$ entitles us to choose any significance, $a$, for $x$ and infer $\phi a$; similarly $\phi a$ entitles us to infer $(\exists x)(\phi x)$.

As well as a new axiom we need a new rule of inference, namely:

$$\frac{p \Rightarrow \phi x}{p \Rightarrow (x)(\phi x)} \tag{2.118}$$

where $p$ is any proposition not containing $x$.

Another useful rule can be derived from the one above viz:

$$\frac{\phi x \Rightarrow p}{(\exists x)(\phi x) \Rightarrow p} \tag{2.119}$$

where $p$ is any proposition not containing $x$.

The proof is:

$$\phi x \Rightarrow p \tag{2.120}$$

Premiss

$$p' \Rightarrow \phi' x \tag{2.121}$$
(2.120) and (1.94: $\phi x$ for $p$, $p$ for $q$)

$$p' \Rightarrow (x)(\phi' x) \tag{2.122}$$
(2.121) and (2.118: $p'$ for $p$, $\phi'$ for $\phi$)

$$(x)'(\phi' x) \Rightarrow p \tag{2.123}$$
(2.122) and (1.94: $p'$ for $p$, $(x)(\phi' x)$ for $q$)

$$(\exists x)(\phi x) \Rightarrow p \tag{2.124}$$
(2.123) and (2.5: Definition)

The first task for which we need the new rule of inference is to establish a result reached intuitively in section **2.2**, i.e.:

$$(y)(\phi y) \equiv (x)(\phi x) \tag{2.125}$$

$$(x)(\phi x) \Rightarrow \phi a \tag{2.126}$$
(2.116: Axiom)

$$(y)(\phi y) \Rightarrow \phi x \tag{2.127}$$
(2.126: $y$ for $x$, $x$ for $a$)

$$(y)(\phi y) \Rightarrow (x)(\phi x) \tag{2.128}$$
(2.127) RI

$$(x)(\phi x) \Rightarrow (y)(\phi y) \tag{2.129}$$
(2.128: $y$ for $x$, $x$ for $y$)

$$(y)(\phi y) \equiv (x)(\phi x) \tag{2.125}$$
(2.128), (2.129) and (1.121)

Notice that in appealing to the rule of inference (2.118) we are justified in taking $(y)(\phi y)$ for $p$, since it does not contain $x$.

We may also use the new rule of inference to establish another result reached intuitively – namely that of (2.11):

$$(x)(\phi x \Rightarrow \psi x) \Rightarrow ((x)(\phi x) \Rightarrow (x)(\psi x)) \tag{2.130}$$

The proof is as follows:

$$(x)(\phi x \Rightarrow \psi x) \Rightarrow (\phi a \Rightarrow \psi a) \tag{2.131}$$
(2.116: $\phi x \Rightarrow \psi x$ for $\phi x$)

$$(x)(\phi x \Rightarrow \psi x) \Rightarrow (\phi' a \vee \psi a) \tag{2.132}$$
(2.131) and (1.2: Definition)

$$\phi' a \Rightarrow (\exists x)(\phi' x) \tag{2.133}$$
(2.117: $\phi'$ for $\phi$)

$$(\phi' a \vee \psi a \Rightarrow ((\exists x)(\phi' x) \vee \psi a) \tag{2.134}$$
RI (1.26) and (2.133)

$$(x)(\phi x \Rightarrow \psi x) \Rightarrow ((\exists x)(\phi' x) \vee \psi a) \tag{2.135}$$
(2.132) and (2.134)

$$((x)(\phi x \Rightarrow \psi x) \ \Rightarrow (\exists x)(\phi' x)) \vee \psi a \tag{2.136}$$

RI (1.18: $p'$ for $p$) and (2.135)

$$((x)(\phi x \Rightarrow \psi x) \Rightarrow (\exists x)(\phi' x))' \Rightarrow \psi a \tag{2.137}$$

(2.136) and (1.69: Definition)

$$((x)(\phi x \Rightarrow \psi x) \Rightarrow (\exists x)(\phi' x))' \Rightarrow (x)(\psi x) \tag{2.138}$$

RI – since $((x)(\phi x \Rightarrow \psi x) \Rightarrow (\exists x)(\phi' x))'$ does not contain $x$

$$((x)(\phi x \Rightarrow \psi x) \Rightarrow (\exists x)(\phi' x)) \vee (x)(\psi x) \tag{2.139}$$

(2.138) and (1.69: Definition)

$$(x)(\phi x \Rightarrow \psi x) \Rightarrow ((\exists x)(\phi' x) \vee (x)(\psi x)) \tag{2.140}$$

RI (1.18) and (2.139)

$$((\exists x)(\phi' x) \vee (x)(\psi x)) \equiv ((x)(\phi x) \Rightarrow (x)(\psi x)) \tag{2.141}$$

(1.2) and (2.5: Definition)

$$(x)(\phi x \Rightarrow \psi x) \Rightarrow ((x)(\phi x) \Rightarrow (x)(\psi x)) \tag{2.130}$$

RI (2.140) and (2.141)

Note that (2.138), which is the crucial step in this proof, depends on the fact that, appearances notwithstanding, $((x)(\phi x \Rightarrow \psi x) \Rightarrow (\exists x)(\phi' x))'$ does not contain $x$. The expressions on both sides of the main sign of implication are 'closed' with respect to $x$; any other symbol might have been used in place of $x$, as (2.125) makes clear (cf. section 2.2(b)).

We also need to be able to infer $(x)(\phi x \wedge \psi x)$ from $(x)(\phi x) \wedge (x)(\psi x)$ and conversely. This was assumed without proof in (2.13):

$$(x)(\phi x) \wedge (x)(\psi x) \tag{2.142}$$

Premiss

$$(x)(\phi x) \Rightarrow \phi y \tag{2.143}$$

(2.116: $y$ for $a$)

$$(x)(\psi x) \Rightarrow \psi y \tag{2.144}$$

(2.116: $\psi$ for $\phi$, $y$ for $a$)

$$((x)(\phi x) \wedge (x)(\psi x)) \Rightarrow (\phi y \wedge \psi y) \tag{2.145}$$

(2.143, (2.144), (1.39) and (1.122)

$$((x)(\phi x) \wedge (x)(\psi x)) \Rightarrow (y)(\phi y \wedge \psi y) \tag{2.146}$$

(2.145) and (2.118: $\phi \wedge \psi$ for $\phi$, $y$ for $x$)

$$(y)(\phi y \wedge \psi y) \tag{2.147}$$

(2.142) and (2.146)

$$(x)(\phi x \wedge \psi x) \tag{2.148}$$

(2.147) and (2.1)

The proof of the converse runs as follows:

$$(x)(\phi x \wedge \psi x) \tag{2.148}$$

Premiss

$$(y)(\phi y \wedge \psi y) \qquad (2.149)$$

(2.148) and (2.1)

$$(y)(\phi y \wedge \psi y) \Rightarrow (\phi x \wedge \psi x) \qquad (2.150)$$

(2.116: $\phi \wedge \psi$ for $\phi$, $y$ for $x$, $x$ for $a$)

$$(\phi x \wedge \psi x) \Rightarrow \phi x \qquad (2.151)$$

(1.6: $\phi x$ for $p$, $\psi x$ for $q$)

$$(y)(\phi y \wedge \psi y) \Rightarrow \phi x \qquad (2.152)$$

(2.150) and (2.151)

$$(y)(\phi y \wedge \psi y) \Rightarrow (x)(\phi x) \qquad (2.153)$$

RI (2.152)

$$(\phi x \wedge \psi x) \Rightarrow \psi x \qquad (2.154)$$

(1.6: $\psi x$ for $p$, $\phi x$ for $q$)

$$(y)(\phi y \wedge \psi y) \Rightarrow \psi x \qquad (2.155)$$

(2.150) and (2.154)

$$(y)(\phi y \wedge \psi y) \Rightarrow (x)(\psi x) \qquad (2.156)$$

RI (2.153)

$$(y)(\phi y \wedge \psi y) \Rightarrow ((x)(\phi x) \wedge (x)(\psi x)) \qquad (2.157)$$

(1.39), (2.154) and (2.156)

$$(x)(\phi x) \Delta (x)(\psi x) \qquad (2.142)$$

(2.149) and (2.157)

The remaining theorems concerning quantifiers, section **2.2**(d), follow without difficulty.

(b) *Russell's antinomy*

Returning to the first question raised in this section we naturally ask ourselves what the matrix $\phi x$ or $\phi x,y$ really denotes. In a sense we should like to be able to keep the question open, substituting $x \in y$ for $\phi x,y$ for example only when we wished to emphasise that whatever significance were given to $x$, $y$ the resulting proposition was atomic, but otherwise contenting ourselves with the statements made in the earlier part of the chapter that $\phi x$ (or $\phi x,y$) is to signify a proposition ('true' for some $x$, $y$ and 'false' for others) whenever a particular meaning is given to $x$, $y$ and defining the set, $\phi$, of just those entities which when substituted for $x$, $y$ render it 'true'.

The trouble is that – to limit ourselves for the moment to a single argument or variable – a set of such $x$ is not so easily defined. For once it be granted that a proposition can take the form $x \in y$ one instance of $\phi x$ would be $x \in x$[17] and another would be $x \notin x$. What happens when we attempt to define $\phi$ in the latter case?

In section 2.3, membership in a set $X$ was defined by means of:

$$x \in X \equiv (i)(T \in i \Rightarrow x \in i) \qquad \text{Df} \qquad (2.44)$$

where $T$ was the subject having just those predicates common to all members of the set and no others. The corresponding definition in the more general case, where the form of the matrix $\phi$ is left indeterminate, would be:

$$x \in \hat{\phi} \equiv \phi x \qquad \text{Df} \tag{2.158}$$

However if we now take $\phi x$ to be $x \notin x$ we are led into a contradiction, for the definition allows us to infer

$$x \in \hat{\phi} \equiv x \notin x \tag{2.159}$$

an equivalence which is to be 'true' for all $x$ whatsoever. Taking $\hat{\phi}$ for $x$ we reach

$$\hat{\phi} \in \hat{\phi} \equiv \hat{\phi} \notin \hat{\phi} \tag{2.160}$$

which is the contradiction referred to.

Generally known as **Russell's antinomy**, this is the direct outcome of trying to include among the entities that can enter into an atomic proposition any and every set, despite the fact that the very definition of some sets precludes them from such membership. The classic illustration is the attempt to define 'the village barber' as the man who shaves every man in the village who does not shave himself. (In this illustration *membership in* is taken as the equivalent of *being shaved by*.) Such a definition excludes the barber from membership of the village, for such membership means that if he doesn't shave himself he does shave himself, and vice versa. The barber being the (unit) set of those who shave the village non-self-shavers we cannot include the barber himself in the set of villagers, which we have already made use of in defining him.

(i) Circular definition

We have so far been able to avoid the possibility of circular definition by using only subjects (i.e., elements in atomic propositions appearing on the left-hand-side of $\in$) to define sets (elements appearing only on the right-hand side). Then, in effect by allowing the sign of inclusion, $\subset$, to replace $\in$, we have developed a calculus in which all the elements are sets (or classes)[18] and the notion of class-membership gives way to class-inclusion.

The trouble with this is that when we wish to form classes of classes (which is to be the basis of our definition of number) it becomes necessary to develop a new calculus in which class inclusion gives way to class–class inclusion and so on indefinitely. This is, in fact, the method developed in *Principia Mathematica* where a proposition of the form $x \in y$ is considered meaningful only if $x$, $y$ denote elements of appropriate types.

However, attempts have been made to overcome the difficulty in other ways and the rest of this section is devoted to a very brief introduction to one such attempt.

(ii) Quine's definition of class-membership

An obvious starting-point is the exclusion from any mention in the definition of class-membership of any element which is inherently non-classifiable (of which the class of those classes which are not members of themselves is only one). This is achieved by means of a definition which ensures that every element that enters into the definition of $\phi$ is itself classifiable (in the event of its being a member of $\hat{\phi}$ in some subset of $\hat{\phi}$) without assuming its membership in $\hat{\phi}$ itself.

The definition runs:

$$x \in \hat{\phi} \equiv (\exists y)((x \in y) \wedge (i)(i \in y \Rightarrow \phi i)) \text{ Df} \qquad (2.161)$$

Special consideration has of course to be given to the case where $\phi$ is a unit set and therefore has no proper subsets. This is done in section **2.5**(c).

(iii) Identity

If we adopt the foregoing definition our first task will be to ensure that under it sets have just the characteristics that they had before. To facilitate this we introduce the notion of **identity**:

$$l = m \equiv (i)(i \in l \equiv i \in m) \qquad \text{Df} \qquad (2.162)$$

$$l \neq m \equiv (l = m)' \qquad \text{Df} \qquad (2.163)$$

$l = m$ is pronounced *l is identical with m*; and an immediate inference from this definition, taken in conjunction with (2.161) is that if $l$ stands for a set, $\hat{\phi}$, $m$ can stand for it also since $x \in l$ can be replaced by $x \in m$ and vice versa.

Other immediate inferences are:

$$l = m \equiv m = l \qquad (2.164)$$

$$(k = l \wedge l = m) \Rightarrow k = m \qquad (2.165)$$

$$l = l \qquad (2.166)$$

(iv) Set membership

Let us now reconsider the definition of (2.44), our original definition of set-membership. We wish to compare membership in a set $X$ as there defined with membership in $\hat{\phi}$ as defined in (2.161) in the particular case where $\phi i$ stands for $(j)(T \in j \Rightarrow i \in j)$.

$$x \in \hat{\phi} \equiv (\exists y)(x \in y \wedge (i)(i \in y \Rightarrow (j)(T \in j \Rightarrow i \in j))) \qquad (2.167)$$

Premiss

$$x \in \hat{\phi} \equiv (\exists y)((x \in y) \wedge (i)(i \in y \Rightarrow i \in X)) \qquad (2.168)$$

(2.167) and (2.44: $j$ for $i$, $i$ for $x$)

$$x \in \hat{\phi} \equiv (\exists y)(x \in y \wedge y \subset X) \qquad (2.169)$$

(2.168) and (2.58) Definition

$$(x \in y \wedge y \subset X) \Rightarrow x \in X \qquad (2.170)$$

(2.70: $y$ for $Y$, $X$ for $Z$)

$$(\exists y)(x \in y \wedge y \subset X) \Rightarrow x \in X \qquad (2.171)$$

RI (2.170)

$$x \in \hat{\phi} \Rightarrow x \in X \qquad (2.172)$$

(2.169) and (2.171)

$$x \in X \Rightarrow (x \in X \wedge X \subset X) \qquad (2.173)$$

(1.34)

$$(x \in X \wedge X \subset X) \Rightarrow (\exists y)(x \in y \wedge y \subset X) \qquad (2.174)$$

(2.117)

$$x \in X \Rightarrow (\exists y)(x \in y \wedge y \subset X) \qquad (2.175)$$

(2.173) and (2.174)

$$x \in X \Rightarrow x \in \hat{\phi} \qquad (2.176)$$

(2.175) and (2.169)

$$x \in X \equiv x \in \hat{\phi} \qquad (2.177)$$

(2.172) and (2.176)

$$(p \vee p') \Rightarrow (x \in X \equiv x \in \hat{\phi}) \qquad (2.178)$$

(1.53: $(p \vee p')$ for $p$, $(x \in X \equiv x \in \hat{\phi})$ for $q$) and (2.177)

$$(p \vee p') \Rightarrow (x)(x \in X \equiv x \in \hat{\phi}) \qquad (2.179)$$

RI (2.178) since $(p \vee p')$ does not contain $x$)

$$p \vee p' \qquad (2.180)$$

(1.3)

$$(x)(x \in X \equiv x \in \hat{\phi}) \qquad (2.181)$$

(2.179) and (2.180)

$$X \equiv \hat{\phi} \qquad (2.182)$$

(2.162: $x$ for $i$, $X$ for $l$, $\hat{\phi}$ for $m$) and (2.181)

We have thus shown that $X$ is identical with $\hat{\phi}$. If instead of defining $\phi i$ by $(j)(T \in j \Rightarrow i \in j)$ we had defined it by $i \in X$ we should of course have reached the same conclusion. Thus, under either definition we have shown that every set is identical with the class of those elements that are its members – a result which may seem trivial until it is recalled that we have as yet no means of identifying members of a class other than by identifying the various sets to which each belongs.

We have also shown that if two classes are equivalent or, more precisely, if the matrices defining them are equivalent for all significations of their arguments then the classes themselves are identical.

It is convenient at this point to generalise the notation introduced on p. 54 by means of the following definition:

$$\{x \mid \phi x\} \equiv \hat{\phi} \qquad \text{Df} \tag{2.183}$$

Where $a, b, c, \ldots,$ stands for a finite[18] list of distinct elements we write:

$$\{a, b, c, \ldots\} \underset{\text{Df}}{\equiv} \{x \mid x \in a \vee x \in b \vee x \in c \vee \ldots\} \tag{2.184}$$

(iv) Meaning of ∈

Expression (2.170) in the proof of (2.182) depended on the substitution of $y$ for $Y$ in (2.70). But there $Y$ symbolised a set. We have not so far assumed that any symbol appearing on the right-hand side of ∈ is necessarily a set in the meaning given to that concept either in (2.44) or (2.161). Suppose $y$ does not stand for some $X$ or some $\hat{\phi}$? In this case it becomes necessary to define $x \in y$ as *x is identical with y*, i.e., as a proposition whose assertion entitles us to substitute $x$ for $y$ or $y$ for $x$.

While in earlier sections, notably **2.2**(a), hints were dropped as to the meaning to be given to ∈ and some intuitive understanding assumed of what is to be understood by membership in a class, nothing would have had to be altered in any of the subsequent proofs if it had been decided to give to ∈ the meaning of 'loves', 'hates', 'is distinct from' or any other connective between nouns.

The definition we have just proposed, however, puts an end to this uncertainty, for it establishes a comparable relationship between symbols appearing on opposite sides of ∈ as did the relation of identity (defined in section **2.5**(b)(iii)) between those appearing on the same – viz. the right-hand side. In fact it will be shown, in section **2.5**(c)(i), that no ambiguity arises if in this case also we write $x = y$ and use this as a definition of identity among members of a class that are not themselves classes.

A little reflection will show that without some such convention in regard to the use of ∈, it would be impossible, while retaining its characteristic property of ordering the parts of a proposition, to allow any symbol to appear on either side of it, since the copula could without ambiguity be treated as part of the predicate (which would then be 'branded' as predicate and would be incapable of assuming the status of subject in some other proposition).

At all events this convention enables us to close the proof of (2.182), since if $y$ does not there stand for a class, then $x$ may be substituted for it and we reach the conclusion $x \in \hat{\phi} \equiv x \in X$ directly.

This understanding of $x \in y$ in the case where $y$ is not a set also enable us to extend (2.164)–(2.167) to the case s where $k, l, m$ are not sets as follows:

$$x = y \equiv y = x \tag{2.185}$$

$$(x = y \wedge y = z) \Rightarrow x = z \tag{2.186}$$

$$x = x \tag{2.187}$$

(c) *The unit class, universal and null classes*

We must now make sure that, in the light of (2.161), we still have a satisfactory definition of the unit class. Our previous definition was:

$$x \in [a] \equiv (i)(a \in i \equiv x \in i) \qquad \text{Df} \tag{2.89}$$

(i) Validity of the definition

There is in fact no need to change this by reason of Russell's antinomy since if $a$ turns out to be non-classifiable it simply tells us that $[a]$ is null (2.221).

However we do need some assurance that on this definition we can infer $x = a$ from $x \in [a]$ and conversely.

Consider first the inference of $x = a$ from $x \in [a]$. If $i$ is not standing for a class the inference is immediate, since $x$ can be substituted for $i$ whenever $a$ can; and, in particular when $i$ takes the significance of $a$.

If $i$ is standing for a class the inference is valid also, since if $a$ were not identical with $x$, it would be possible to define a class $\phi$ of which $a$ was a member and $x$ was not, thus negating the premiss. The proof is as follows:

$$(i)(a \in i \equiv x \in i) \tag{2.188}$$

Premiss

$$j \in \hat{\phi} \equiv (j \in [a]\ \hat{\phi} j \neq x) \tag{2.189}$$

Definition

$$(j)(a \in j \equiv a \in j) \equiv a \in [a] \tag{2.190}$$

(2.89: $a$ for $x$, $j$ for $i$)

$$(i)(a \in i \equiv x \in i) \Rightarrow (a \in j \equiv a \in j) \tag{2.191}$$

(1.51: $(i)(a \in i \equiv x \in i)$ for $p$, $a \in j$ for $q$)

$$(i)(a \in i \equiv x \in i) \Rightarrow (j)(a \in j \equiv a \in j) \tag{2.192}$$

(2.118: $j$ for $x$, $(a \in j \equiv a \in j)$ for $\phi x$) and (2.191)

$$(i)(a \in i \equiv x \in i) \Rightarrow a \in [a] \tag{2.193}$$

(2.190) and (2.192)

$$a \in [a] \tag{2.194}$$

(2.188) and (2.193)

$$a \in \hat{\phi} \equiv (a \in [a] \wedge a \neq x) \tag{2.195}$$

(2.189: $a$ for $j$)

$$(a \in [a] \wedge a \neq x) \Rightarrow a \in \phi \quad (2.196)$$

(1.121) and (2.195)

$$((a \in [a] \wedge a \neq x) \Rightarrow a \in \phi) \Rightarrow (a \in [a] \Rightarrow ((a \neq x) \Rightarrow a \in \phi))) \quad (2.197)$$

(1.44: $a \in [a]$ for $p$, $a \neq x$ for $q$, $a \in \phi$ for $r$)

$$a \in [a] \Rightarrow (a \neq x) \Rightarrow a \in \phi) \quad (2.198)$$

(2.196) and (2.197)

$$a \neq x \Rightarrow a \in \phi \quad (2.199)$$

(2.194) and (2.198)

$$(i)(a \in i \equiv x \in i) \Rightarrow (a \in \phi \equiv x \in \phi) \quad (2.200)$$

(2.116: $i$ for $x$, $a \in i \equiv x \in i$ for $\phi x$)

$$a \in \phi \equiv x \in \phi \quad (2.201)$$

(2.188) and (2.200)

$$a \neq x \Rightarrow x \in \phi \quad (2.202)$$

(2.199) and (2.201)

$$x \notin \phi \Rightarrow a = x \quad (2.203)$$

(2.202) and (1.15: $a \neq x$ for $p$, $x \in \phi$ for $q$)

$$x \notin \phi \equiv (x \notin [a] \vee x = x) \quad (2.204)$$

(2.189: $x$ for $j$) and (1.11)

$$(x \notin [a] \vee x = x) \Rightarrow x \in \phi \quad (2.205)$$

(2.204) and (1.121)

$$((x \notin [a] \vee x = x) \Rightarrow x \in \phi) \Rightarrow (x = x \Rightarrow x \notin \phi) \quad (2.206)$$

(1.52: $x \in [a]$ for $p$, $x = x$ for $q$, $x \in \phi$ for $r$)

$$x = x \Rightarrow x \notin \phi \quad (2.207)$$

(2.205) and (2.206)

$$x = x \quad (2.208)$$

(2.187)

$$x \notin \phi \quad (2.209)$$

(2.207) and (2.208)

$$a = x \quad (2.210)$$

(2.203) and (2.209)

$$a = x \equiv x = a \quad (2.211)$$

(2.185: $a$ for $y$)

$$x = a \quad (2.212)$$

(2.210) and (2.211)

(ii) Inference of $x \in [a]$ from $x = a$

As to the inference of $x \in [a]$ from $x = a$, the following will serve:

$$x = a \quad (2.212)$$

Premiss

$$x = a \Rightarrow (a \in i \equiv a \in i) \qquad (2.213)$$

(1.51: $x = a$ for $p$, $a \in i$ for $q$)

$$x = a \Rightarrow (x \in i \equiv a \in i) \qquad (2.214)$$

(2.212) and (2.213)

$$x = a \Rightarrow (i)(x \in i \equiv a \in i) \qquad (2.215)$$

(2.118) and (2.214)

$$(i)(x \in i \equiv a \in i) \qquad (2.216)$$

(2.212) and (2.215)

$$x \in [a] \equiv (i)(x \in i \equiv a \in i) \qquad (2.217)$$

(2.89)

$$x \in [a] \qquad (2.218)$$

(2.216) and (2.217)

(iii) Unit class

It is to be noted that nowhere in the foregoing have we had to identify a unit class, $[a]$, with its own sole member, $a$; and in general this identification is not possible. However in the case where $a$ is known to be a non-class this identification follows from (2.89), since, for all $i$,

$$i \in a \equiv i = a; \qquad \text{and } i = a \equiv i \in [a] \qquad (2.219)$$

The formal proof is left as an exercise.

(iv) Universal and null classes

We conclude this sub-section by observing that the definition of identity and the inference $x = x$ afford us the opportunity of more general definitions of the **universal** and **null classes** than were possible in (2.66) and (2.67), namely:

$$\vee \equiv \{x \mid x = x\} \qquad \text{Df} \qquad (2.220)$$
$$\wedge \equiv \{x \mid x \neq x\} \qquad \text{Df} \qquad (2.221)$$

(d) *Classes of classes*

Can classes qualify for class-membership? If so are the classes of which they are members a distinct genre? These are the questions with which this section has been chiefly concerned. We have not pressed them to a conclusion; to do so would be beyond our scope. The answers would seem to be 'yes' and 'no', respectively, provided we are concerned only with finite sets; and provided also that 'non-sets', 'individuals' (call them what you will) are always identified with their own unit sets.

What we have shown is that **classes** – defined by means of propositional matrices – identify with sets – defined by means of a typical member. It follows that classes thus defined can be complemented, alternated and conjoined in the same way as was demonstrated in the case of sets in sections **2.2–2.4**.

Thus from:

$$x \in \hat{\phi} \equiv (\exists y)((x \in y) \wedge (i)(i \in y \Rightarrow \phi i)$$

and
$$x \in \hat{\psi} \equiv (\exists y)((x \in y) \wedge (i)(i \in y \Rightarrow \psi i)$$

we can infer

$$x \notin \hat{\phi} \equiv (\exists y)((x \in y) \wedge (i)(i \in y \Rightarrow \phi' i)$$

$$x \in \hat{\phi} \vee x \in \hat{\psi} \equiv (\exists y)((x \in y) \wedge (i)(i \in y \Rightarrow (\phi i \vee \psi i))$$

and
$$x \in \hat{\phi} \wedge x \in \hat{\psi} \equiv (\exists y)((x \in y) \wedge (i)(i \in y \Rightarrow (\phi i \wedge \psi i))$$

---

**Exercises**

2.21 Put the following propositions into logical form, explaining the symbolism used, and comment on the logical relations that exist between them:
(a) Not only those who appreciate Dickens' novels admire his intentions in writing them.
(b) There are some people who neither admire Dickens' intentions in writing his novels nor appreciate the novels themselves.
(c) Nobody who appreciates Dickens' novels fails to admire his intentions in writing them.
(d) There is no universal correlation between appreciating Dickens' novels and admiring his intentions in writing them. (London)

2.22 If $A \equiv \{1, 2, \{1, 2\}, 3\}$; $B \equiv \{1, 2, 3\}$; $C \equiv \{1, 2\}$; which of the following is false?
(a) $C \in A$
(b) $C \subset A$
(c) $C \in B$
(d) $C \subset B$ (Edinburgh)

2.23 If $\phi x,y$ is a propositional matrix involving two arguments and $p \equiv (x)((\exists y)(\phi x,y))$; $q \equiv (\exists y)((x)(\phi x,y))$, which of the following can be inferred?
(a) $p \equiv q$
(b) $p \Rightarrow q$
(c) $q \Rightarrow p$ (Edinburgh)

2.24 A set, $A$, of points in a plane is called open if, given any member $P$ of $A$, there exists a positive number, $\delta$, such that all points whose distance from $P$ is less than $\delta$ are also members of $A$; $A$ is called closed if the set of points not members of $A$ is open. $P$ is called a limit point of $A$ if, given any positive number $\delta$, there exists a member of $A$ other than $P$ whose distance from $P$ is less than $\delta$.

Show that $A$ is closed if and only if all its limit points are members of it. (Cambridge: adapted)

2.25 Render the following argument into symbols and determine its validity:
Although the Tate gallery possesses some of Picasso's greatest pictures it does not have one painted before he was fifty. As he hasn't produced a great painting since 1939 he must have been born before 1890.
(London)

2.26 Discuss the following argument for scepticism:
To justify any proposition we must produce its premisses. To justify those premisses we must produce their premisses. Therefore, either an infinite regress or the fact that we must start from arbitrary, unjustifiable beliefs must be admitted. Therefore there can be no such thing as a wholly rational belief. (Cambridge)

---

**Notes on chapter 2**

1. In fact the hypothesis is seldom stated in quite such a general form; it does not hold, for example, for very large or very small volumes.
2. Witkowski's measurements, quoted in Glazebrook, *Dictionary of Applied Physics,* (1922), Vol. 1, p. 889, are here taken and rendered into SI units to an accuracy of three significant figures.
3. Not all quantified propositions are free of this restriction, obviously. The proposition *all fifty-pence pieces have fourteen edges* is one which, true or false, conjoins only a finite number of constituent propositions.
4. Some writers treat the copula as existing only as part of the predicate which then includes the words *is* or *is a*. But this makes it difficult to use the same symbol on both sides of the copula. See section 2.5(b)(iv), page 69).
5. Where $x$ is governed by a quantifier it is said to be a **bound variable**; otherwise a **free variable.**
6. It is unfortunate that in symbolic logic the word 'argument' has two senses. For a sequence of inferences we can perhaps, though loosely, use the word 'demonstration'.
7. From this point onwards the inference:

$$\begin{array}{c} p \equiv q \\ q \equiv r \\ \hline p \equiv r \end{array}$$

is assumed without proof and written:

$$\begin{array}{r} p \equiv q \\ \equiv r \end{array}$$

8. It may be said that the 'typical member' of a set is a logical fiction; but then so is the set which it is used to define. No-one has ever seen a set – only an aggregate!
9. After *John Venn*, author of *Symbolic Logic*, 1881.
10. Again, convention is unfortunate. Since $\cup$ corresponds to $\vee$ and $\cap$ to $\wedge$, it would have seemed more sensible to make $\supset$ or $\rightarrow$ correspond to $\Rightarrow$.
11. The enunciation of the formal rules of inference required is postponed to (2.118).
12. The **converse** of $p \Rightarrow q$ is $q \Rightarrow p$.
13. The premiss here referred to is in fact the axiom of quantification whose formal enunciation is postponed to (2.116).
14. In *Symbolic Logic*, 1896.
15. In the original this is expressed as 'No-one is despised who can manage a crocodile'.
16. *a* here stands for any symbol that can take the place of $x$.
17. The notion of 'membership-in-oneself' is not inherently self-contradictory. 'Polysyllabicity' may be defined as the predicate common to just those words that have more than one syllable – a predicate which it itself enjoys. Similarly 'monosyllabicity' is a predicate which does not enjoy itself.
18. The words 'set' and 'class' are generally treated as synonyms. In this chapter we use 'set' when defining by means of a typical member; 'class' when defining by means of a propositional matrix. The two usages are reconciled in section **2.5**(c)(iv). A slightly different convention is then adopted in chapter 3.

## Definitions used in chapter 2

(8) $\phi' x$ for $(\phi x)'$
(9) $(x)'(\phi x)$ for $((x)(\phi x))'$
(10) $(\exists x)(\phi x)$ for $(x)'(\phi' x)$
(11) $(\exists x)'(\phi x)$ for $(x)(\phi' x)$
(12) $x \in \hat{\phi}$ for $(\exists y)(x \in y \wedge (i)(i \in y \Rightarrow \phi y))$
(13) $x \notin X$ for $(x \in X)'$
(14) $x \in X'$ for $x \notin X$
(15) $x \in X \cup Y$ for $x \in X \vee x \in Y$
(16) $x \in X \cap Y$ for $x \in X \wedge x \in Y$
(17) $X \subset Y$ for $(x)(x \in X \Rightarrow x \in Y)$
(18) $Y \supset X$ for $X \subset Y$
(19) $X \equiv Y$ for $(x)(x \in X \equiv x \in Y)$
(20) $x \in [a]$ for $(i)(a \in i \equiv x \in i)$
(21) $l = m$ for $(i)(i \in l \equiv i \in m)$

(22) $l \neq m$ for $(l = m)'$
(23) $\{x \mid \phi x\}$ for $\hat{\phi}$
(24) $\{a, b, c, \ldots\}$ for $\{x \mid x \in a \vee x \in b \vee x \in c \ldots\}$
(25) $\mathsf{V}$ for $\{x \mid x = x\}$
(26) $\Lambda$ for $\{x \mid x \neq x\}$

# 3

# THE LOGIC OF RELATIONS

The weakness of a logical system which goes no further into the structure of language than the relation of subject to predicate is that it is unable to deal with quantity. What is characteristic of a quantity is that it can be related in a wide variety of ways to other quantities whereas classes can only be related by inclusion, exclusion (that is, inclusion in the complementary class) or a relation which is neither of these, namely overlapping.

Since relationship lies at the root of the idea of quantity itself we try in this chapter to expand the notion of a proposition as something which relates two parts by a single connective to meet the demands of a wider variety of relations. No new axioms or rules of inference are required and very little in the way of new symbolism. The opportunity is thus afforded of writing out proofs – the general lines of which are now familiar – in a rather shorter form. In particular the syllogism (section 1.4(a)(ii)) is written as:

$$\begin{array}{l} p \Rightarrow q \\ \underline{\quad \Rightarrow r} \\ p \Rightarrow r \end{array}$$

## 3.1 Cause and effect: Mendel's law

One of the most confused and confusing of the ideas which are still current in popular science is that of the relation of 'cause' to 'effect'. Current in a wire is often said to 'cause' the movement of iron filings in its neighbourhood, or gravity to 'cause' the falling of a stone. But it is significant that many pioneers in the field of scientific discovery did not bother themselves with 'causes'. Newton's *hypotheses non fingo* is a celebrated instance. Another is that of Gregor Mendel, whose discovery of the relation between the characteristics of parent and offspring laid the foundation of modern genetic theory. The significance of this discovery, which was linked to no theory of 'causes', has in no way been diminished by subsequent controversy as to the nature of genes.

Mendel's experiments involved the cross-fertilisation of two pure strains of garden pea, one having a smooth surface (the so-called 'recessive' characteristic) and the other a crinkled surface (the so-called 'dominant'). In the second generation after interbreeding began the number of peas displaying the dominant characteristic stood to the number displaying the recessive in a proportion of almost exactly three to one. The larger the number of plants used in the experiment the more exact the proportion. Mendel showed that these results and those obtained from a single generation were consistent with a particulate theory of inheritance – the hypothesis of two factors or 'causes' of inheritance in each individual which were equally mixed in its offspring – the dominant factor accounting for the characteristic of all offspring except those in which both factors were recessive. As to the nature of these 'causes' nothing was and nothing needed to be said.

The logic of Mendel's experiment is even simpler than that which underlies Boyle's Law (section **2.1**). Apart from the precautions necessary to ensure against external fertilisation only three procedures are required:

(i) classifying the offspring according to the characteristics displayed by their parents or grandparents;
(ii) counting the number in each class (which, from the logical point of view, is a much simpler procedure than measuring); and
(iii) comparing the numbers in each class.

In the example cited – the first of many experiments which Mendel undertook – the procedure is still further simplified by the fact that the offspring can fall only into one of two classes – smooth-surfaced and crinkly-surfaced – and these are precisely the classes into which the parents have already been divided. It becomes in this case, therefore a question of relating each offspring to a representative parent or grandparent and counting the number of relations of each type. If $y$ stands for the representative parent chosen and $C$ for the class of pea (crinkly-surfaced, say) to which it belongs, so that $y \in C$, we wish to know for how many $x$'s $x \in C$ may be asserted, where $x$ may stand for any member of the class of offspring.

It is rather simpler however to express this by saying that we wish to know how many $x$'s bear the relationship to $y$ of 'belonging to the same class' (in respect of skin-type). An analogous problem would be that of determining the sex of human offspring in relation to that of a representative parent, where again there are only two possibilities to be considered. In this case we wish to know how many offspring bear the relationship of son to their parents and how many bear the relationship of daughter.

## 3.2 Two-termed relations

### (a) *Ordered pairs*

From the previous paragraph it appears that the establishment of a relationship between two entities is closely connected with the classification of those entities in the same or related classes. In fact, provided that the order of the two entities is preserved the proposition of a relationship, $R$, between two entities $x$ and $y$ – written for convenience as $xRy$ – is equivalent to the proposition that the **ordered pair**[1] $x,y$ belongs to a certain class of ordered pairs, namely that of all pairs between which this relationship[2] holds.

Suppose, for example, that in the instance cited, the universe of discourse is the class of all peas under consideration, whether parents or offspring. Suppose, further, that $\phi y$ stands for *y is a parent,* $\psi y$ for *y is an offspring*; and $\chi y$ for *y is crinkly-surfaced.* Then, since we are concerned only with second-generation offspring, i.e., offspring that are not themselves parents, the class $\chi$ of crinkly-surfaced, second-generation offspring will be defined by:

$$x \in x \equiv \psi x \wedge \phi' x \wedge \chi x$$

The relationship, $xRy$, obtaining between $x$ and $y$ when $x$ is second-generation, $y$ is original and both are crinkly-surfaced, will be defined by:

$$xRy \equiv \psi x \wedge \phi' x \wedge \chi x \wedge \psi' y \wedge \phi y \wedge \chi y \tag{3.1}$$

and that obtaining when $x$ is second generation, $y$ is original and $y$ is crinkly-surfaced will be defined by:

$$xSy \equiv \psi x \wedge \phi' x \wedge \psi' y \wedge \phi y \wedge \chi y \tag{3.2}$$

Since, given the classes defined by $\phi$, $\psi$ and $\chi$ we can in practice decide easily enough which of the elements of the concourse are members of them we can also determine which of the ordered pairs $x, y$ taken from all these elements belong to the classes defined in (3.1) and (3.2).

A difficulty may however occur to the reader at this point, namely that an expression of the kind $\psi x \wedge \chi y$ does not seem to define a class of ordered pairs in quite the same way that, say, $\phi x$ defines a class of individuals. In the definition of the latter (2.161) the variable $x$ occurs in such a way that the substitution for it of any particular element, $a$ or $b$, presents no difficulty. But in defining a class of ordered pairs it is not necessarily the case that the pair occurs as a pair in the definition so that substitution of a particular pair $a,b$ for $x,y$ is not simple but involves a double substitution – of $a$ for $x$ and $b$ for $y$.

Now the substitution of one element, $a$, for another element, $x$ in chapter 2 rested on one of two claims. Either $x$ was a variable of quantification and $a$ was a particular instance of it, or $x$ and $a$ were identical elements. In the first case our difficulty is met by an extension of the notion of quantification to cover more than one[3] variable by means of the following definition:

$$(x,y)(\phi x,y) \equiv (x)((y)(\phi x,y)) \quad \text{Df} \qquad (3.3)$$

where $\phi x,y$ is a matrix in which $x,y$ serve as independent placekeepers and do not necessarily occur as a pair.

By means of this definition we can readily obtain an assertion to correspond to (2.116) and in inference to correspond to (2.118) as follows:

$$(x,y)(\phi x,y) \Rightarrow (x)((y)(\phi x,y)) \qquad (3.4)$$
(3.3) and (1.3)

$$\Rightarrow (y)(\phi a,y) \qquad (3.5)$$
(2.116)

$$\Rightarrow \phi a,b \qquad (3.6)$$
(2.116: $b$ for $a$)

$$(x,y)(\phi x,y) \Rightarrow \phi a,b \qquad (3.7)$$
(3.4), (3.5) and (3.6)

$$p \Rightarrow \phi x,y \qquad (3.8)$$
Premiss

$$p \Rightarrow (y)(\phi x,y) \qquad (3.9)$$
(3.8) and (2.118: $y$ for $x$)

$$p \Rightarrow (x)((y)(\phi x,y)) \qquad (3.10)$$
(3.9) and (2.118)

$$(x)((y)(\phi x,y)) \Rightarrow (x,y)(\phi x,y) \qquad (3.11)$$
(3.3) and (1.3)

$$p \Rightarrow (x,y)(\phi x,y)$$
(3.10) and (3.11)

In the case where substitution rests upon identity we have again to extend our definition to cover the identity of two ordered pairs.

$$x,y = u,v \equiv (x = u \wedge y = v) \quad \text{Df} \qquad (3.12)$$

This justifies the distinct but simultaneous substitution of $a$ for $x$ and $b$ for $y$ as equivalent to the substitution of the pair $a,b$ for the pair $x,y$.

It would now be possible to define a two-termed relation (for example that of parent to child) as a class of ordered pairs and a three-termed relation (for example that of adjudicator between $x$ and $y$) as a class of ordered triads and so on. However it is in the long run simpler to regard classes as hitherto defined as a special kind of relation, i.e., a one-termed or monadic relation. In section **2.5** care was taken not to restrict the argument of the formula $\phi$ to a single element, and no part of the argument set out there depends upon that restriction being made. The whole section could in fact be rewritten with the element $x$ replaced by an ordered aggregate of elements $x,y,z,\ldots$, subject only to the necessary extension of the concepts of quantification and identity outlined above.

Treating classes as monadic relations affords us a further advantage. We can now take over the entire symbolism of classes (exclusion, inclusion, identity, etc.) and apply it to relations generally, using capital letters[4] to denote relations as well as classes. In order to clarify this procedure we shall speak of monadic relations as **sets,** and retain the word **class** to cover relations of any order, monadic, dyadic, triadic, etc. The word **relation,** unless otherwise specified will be restricted to **dyadic** or **two-termed relations** and it is with relations in this sense that this chapter is chiefly concerned.

If $R$ is a dyadic relation we adopt the following definition:

$$xRy \equiv x,y \in R \qquad \text{Df} \qquad (3.13)$$

(b) *Summary of relations so far encountered*

(i) The relation of identity between individuals: $l = m$

This was discussed in section **2.5** and the following conclusions stated:

$$l = m \equiv m = l$$
$$(R = l \land l = m) \Rightarrow k = m$$
$$l = l$$

If you have omitted consideration of section **2.5**, you can safely treat identity as an intuitive concept entitling you to replace any symbol by one with which it is identical.

If it becomes necessary to stress the fact of relationship we can replace the sign of identity by $I$ and adopt the formal definition:

$$l,m \in I \equiv (\exists R)(lRm \land (i,j)(iRj \Rightarrow i = j)) \quad \text{Df} \qquad (3.14)$$

cf. (2.161)

(ii) The relation of diversity between individuals: $l \neq m$

This is the complement of identity. Denoting it by $J$ we have

$$l,m \in J \equiv l,m \notin I \qquad \text{Df} \qquad (3.15)$$

(iii) The relation of equivalence between classes

One of the results of section **2.5** was to do away with the distinction between individuals and classes by allowing classes to be formed, having sets as members, and replacing every individual by its own unit set.

Thus equivalence itself becomes the equivalent of identity and:

$$P \equiv Q \text{ may be written as } PIQ$$
$$\text{and } P \not\equiv Q \text{ as } PJQ \qquad (3.16)$$

Here $I$, $J$ are classes of ordered pairs of classes and thus in Russell and Whitehead's terminology of a different type from $P,Q$.

(iv) The relations of inclusion: $P \subset Q$ and $P \supset Q$

These are written as $PKQ$ and $QKP$; the order is of course significant.

(v) The relations of exclusion and overlapping:

$$P \cap Q \equiv \Lambda \quad \text{and} \quad P \cap Q \not\equiv \Lambda \qquad (3.17)$$

These are written as $PLQ$ and $PL'Q$.

(c) *Inverse and complementary relations*

The last two paragraphs afford instances of two types of relations between relations which we now examine more closely.

(i) Complement of a relation

If two relations $R$, $S$ are such that:

$$(x,y)(xRy \not\equiv xSy) \qquad (3.18)$$

then $S$ is said to be the **complement**[6] of $R$ and we write $S \equiv R'$. Formally:

$$x,y \in R' = x,y \notin R \qquad \text{Df} \qquad (3.19)$$

It is evident that $(R')' \equiv R$, so that if $S$ is the complement of $R$, $R$ is the complement of $S$.

(ii) Inverse of a relation

If two relations $R$, $S$ are such that

$$(x,y)(xRy \equiv ySx) \qquad (3.20)$$

then $S$ is said to be the inverse of $R$ and we write $S \equiv R^{-1}$. Formally:

$$x,y \in R^{-1} \equiv y,x \in R \qquad \text{Df} \qquad (3.21)$$

It is evident that $(R^{-1})^{-1} \equiv R$. The proof, which involves an appeal to (3.12) is left as an exercise.

*Worked example*

The complement of the inverse is the inverse of the complement.

$$x,y \in (R^{-1})' \equiv (xR^{-1}y)' \tag{3.22}$$

(3.13) and (3.19)

$$\equiv (yRx)' \tag{3.23}$$

(3.21)

$$\equiv (yR'x) \tag{3.24}$$

(3.19)

$$\equiv y,x \in R' \tag{3.25}$$

(3.12)

$$x,y \in (R^{-1})' \equiv x,y \in (R')^{-1} \tag{3.26}$$

(3.21)

$$(p \vee p') \Rightarrow (x,y \in (R^{-1})' \equiv x,y \in (R')^{-1} \tag{3.27}$$

(3.26) and (1.53)

$$(p \vee p') \Rightarrow (x,y)(x,y \in (R^{-1})' \equiv x,y \in (R')^{-1} \tag{3.28}$$

(3.27) and (3.8)

$$(p \vee p') \Rightarrow ((R^{-1})' \equiv (R')^{-1}) \tag{3.29}$$

(3.28), (2.162) and (3.16)

$$p \vee p' \tag{3.30}$$

(1.3)

$$(R^{-1})' \equiv (R')^{-1} \tag{3.31}$$

(3.29) and (3.30)

(d) *Null and universal relations*

It was observed in section **3.2**(a) that the calculus of classes developed in chapter 2 can be extended to deal with relations of any order. It is not intended to labour this point. Section **3.4** will provide sufficient instances of argument in the calculus of relations to make the matter clear. However the definitions of the null and universal relation are worthy of special mention. Each is the complement of the other and the definitions are:

$$x,y \in \Lambda \equiv x,y \neq x,y \quad \text{Df} \tag{3.32}$$

$$x,y \in V \equiv x,y = x,y \quad \text{Df} \tag{3.33}$$

(e) *Cartesian product of two sets*

If $P$, $Q$ are any two sets, the relation which holds between any member of $P$ and any member of $Q$ and between no other pair of entities is known as the **Cartesian product** of $P$ and $Q$ and is written $P \times Q$:

$$x, y \in P \times Q \equiv (x \in P \wedge y \in Q) \quad \text{Df} \tag{3.34}$$

As an almost immediate inference:

$$(P \times Q)^{-1} \equiv Q \times P \tag{3.35}$$

For example, the Cartesian product of all higher-case letters and all lower-case letters is equivalent to the relation 'being of higher-case than'.

(f) *Relative product of two relations*

If $R$, $S$ are only two relations, the relations which obtains between any $x$ and any $y$ when a $z$ can be found which bears the relation $R$ to $x$ while $y$ bears the relation $S$ to it is termed the **relative product** of $R$ and $S$, and is written $R \mid S$.

$$x,y \in R \mid S \equiv (\exists z)(xRz \wedge zSy) \tag{3.36}$$

Instance: the relative product of 'being the brother of' and 'being the parent of' is equivalent to the relation 'being the uncle of'.

*Worked example*

$$R \mid S \equiv (S^{-1} \mid R^{-1})^{-1} \tag{3.37}$$

$$x,y \in (S^{-1} \mid R^{-1})^{-1} \equiv y, x \in S^{-1} \mid R^{-1} \tag{3.38}$$

(3.21)

$$\equiv (\exists z) yS^{-1}z \wedge zR^{-1}x) \tag{3.39}$$

(3.36)

$$\equiv (\exists z)(zSy \wedge xRz) \tag{3.40}$$

(3.21)

$$\equiv (\exists z)(xRz \wedge zSy) \tag{3.41}$$

(1.8)

$$\equiv x,y \in R \mid S \tag{3.42}$$

The remainder of the proof follows the lines of (3.27)–(3.31).

*Worked example*

$$(R \mid S) \mid T \equiv R \mid (S \mid T)$$

We need two Lemmas.

Lemma 1

$$(x)((y)(\phi x,y)) \equiv (y)((x)(\phi x,y)) \tag{3.43}$$

$$(x)((y)(\phi x,y)) \Rightarrow (y)(\phi a,y) \tag{3.44}$$

(2.116)

$$\Rightarrow \phi a,b \tag{3.45}$$

(2.116)

$$\Rightarrow (a)(\phi a,b) \tag{3.46}$$

(2.118)

$$\Rightarrow (b)((a)(\phi a,b)) \tag{3.47}$$

(2.118)

$$(a)(\phi a,b) \equiv (x)(\phi x,b) \tag{3.48}$$

(2.125)

$$(b)((a)(\phi a,b) \equiv (y)((x)(\phi x,y) \qquad (3.49)$$

(2.125) and (3.48)

$$(x)((y)(\phi x,y) \Rightarrow (y)((x)(\phi x,y) \qquad (3.43)$$

(3.44), (3.45), (3.46), (3.47) and (3.49)

This proves the first part of the Lemma.

$$\psi y,x \equiv \phi x,y \quad \text{Df} \qquad (3.50)$$

$$(y)((x)(\psi x,y)) \Rightarrow (x)((y)(\psi x,y)) \qquad (3.51)$$

(3.50) and (3.43: $y$ for $x$ and $x$ for $y$)

$$(y)((x)(\phi x,y)) \Rightarrow (x)((y)(\phi x,y)) \qquad (3.52)$$

(3.51: $\phi$ for $\psi$)

$$(x)((y)(\phi x,y)) \equiv (y)((x)(\phi x,y)) \qquad (3.53)$$

(3.43) and (3.52)

Lemma 2

$$(\exists x)((\exists y)(\phi x,y)) \equiv (\exists y)((\exists x)(\phi x,y)) \qquad (3.54)$$

$$(\exists x)((\exists y)(\phi x,y) \equiv (x)'((y)(\phi' x,y)) \qquad (3.55)$$

(2.5) and (2.6)

$$\equiv (y)'((x)(\phi' x,y) \qquad (3.56)$$

(3.53: $\phi'$ for $\phi$)

$$\equiv (\exists y)((\exists x)(\phi x,y)) \qquad (3.57)$$

(2.5) and (2.6)

$$(\exists x)((\exists y)(\phi x,y)) \equiv (\exists y)((\exists x)(\phi x,y)) \qquad (3.54)$$

(3.55), (3.56) and (3.57)

Theorem

$$x,y \in R \mid (S \mid T) \equiv (\exists z)(xRz \wedge zS \mid Ty) \qquad (3.58)$$

(3.36)

$$\equiv (\exists z)(xRz \wedge (\exists w)(zSw \wedge wTy) \qquad (3.59)$$

(3.36)

$$\equiv (\exists z)((\exists w)(xRz \wedge (zSw \wedge wTy))) \qquad (3.60)$$

(2.42)

$$\equiv (\exists z)((\exists w)((xRz \wedge zSw) \wedge wTy)) \qquad (3.61)$$

(1.17)

$$\equiv (\exists w)((\exists z)((xRz \wedge zSw) \wedge wTy)) \qquad (3.62)$$

(3.54)

$$\equiv (\exists w)((\exists z)(xRz \wedge zSw) \wedge wTy) \qquad (3.63)$$

(2.42)

$$\equiv (\exists w)(xR \mid Sw \wedge wTy) \qquad (3.64)$$

(3.36)

$$\equiv x,y \in (R \mid S) \mid T \quad (3.65)$$
(3.36)

The rest of the proof as in (3.27)–(3.31).

(g) *Domain and counter-domain*

It is characteristic of any two-termed relation to have three sets, not necessarily distinct, associated with it, namely the set of all possible first terms, referred to as the **domain** of the relation; the set of all possible second terms referred to as the **counter-domain**; and the union of domain and counter-domain, referred to as the **field**. For any relation, $R$, these sets are denoted by $D(R)$, $C(R)$ and $F(R)$.

$$x \in D(R) \equiv (\exists y)(xRy) \quad \text{Df} \quad (3.66)$$
$$x \in C(R) \equiv (\exists y)(yRx) \quad \text{Df} \quad (3.67)$$
$$x \in F(R) \equiv x \in C(R) \cup D(R) \quad \text{Df} \quad (3.68)$$

For example:

(i) *Being the king of* has as its domain the set of all male hereditary[6] monarchs; and as its counter-domain the set of all kingdoms.

(ii) *Being the spouse of* has as its domain, counter-domain and field the union of the sets of all husbands and wives, i.e., the set of all married persons.

(iii) $P \times Q$ has as its domain $P$; as its counter-domain $Q$; and as its field $P \cup Q$.

---

**Exercises**

3.1 What, in the universe of natural numbers (positive integers and zero), are the domains of *being less than, being greater than, being equal to, being the square root of?*

3.2 What, in the universe of human beings (alive or dead), is the converse of the relation of parenthood; and what is the field of this relation?

3.3 Show that the field of any relation is the field of its inverse. What is the field of the universal relation, and of the relation of diversity in a universe having only one member?

3.4 In (3.22)–(3.31) it was shown that the inverse of the complement is the complement of the inverse. What is this relation in the case where the original relation is the cartesian product of a set with itself? Illustrate this by forming the cartesian product of the set of only children with itself. Take the set of all children as the universe of discourse.

3.5 Show that

(a) $$V^{-1} \equiv V$$

(b) $$\Lambda^{-1} \equiv \Lambda$$

(c) $$(P \times Q)^{-1} \equiv Q \times P$$

3.6 Show that if $A, B, C$ are non-null sets:

$$(A \times B) | (B \times C) \equiv A \times C$$

Does the equivalence hold if either $B$ or $C$ is null?

3.7 Show that $[x,y] \equiv [x] \times [y]$ [7]

---

## 3.3 The calculus of two termed relations

The purpose of this section is simply to familiarise you with the idea that relations can be handled in exactly the same way as sets by providing a few examples of the relations that obtain between them. It can be omitted without loss of continuity.

### (a) *Basic tautologies*

We take as proven those basic tautologies which derive from the definition of a relation as a class of ordered pairs, e.g.:

$$(R \cap S)' \equiv R' \cup S' \qquad (3.69)$$

$$(R \cup S)' \equiv R' \cap S' \qquad (3.70)$$

As in chapter 2 these are assumed listed as in section **1.3(d)**. We can proceed at once to the definitions of **inclusion, identity** and **exclusion.**

### (b) *Inclusion, identity and exclusion among relations*

(i) Inclusion

The inclusion of one relation within another is defined in the obvious manner:

$$R \subset S \equiv (x,y)(x,y \in R \Rightarrow x,y \in S) \text{ Df} \qquad (3.71)$$

For example, *being the son of* is included in *being the child of.*

(ii) Identity

The identity of two relations is similarly defined.

$$(R \equiv S) \equiv (R \subset S \wedge S \subset R) \qquad \text{Df} \qquad (3.72)$$

For example, *being the son of* is identical with *being the male child of.*

The proof that two relations are identical if they have identical ordered pairs as members is exactly analogous to that of (2.182).

(iii) Worked example

The relation of identity is included in the relation of inclusion.

$$(X \subset Y)' \vee (Y \subset X)' \vee (X \subset Y) \qquad (3.73)$$

(1.33)

$$(X \equiv Y)' \vee (X \subset Y) \qquad (3.74)$$

(3.71) and (3.69)

$$(X \equiv Y) \Rightarrow X \subset Y \qquad (3.75)$$

(2.64)

$$X,Y \in I \Rightarrow X,Y \in K \qquad (3.76)$$

Sections 3.2(b)(iii) and (iv)

The rest of the proof as in (3.27)–(3.31).

(iv) Exclusion

Two relations are said to exclude one another when their intersection is null:

$$RLS \equiv (R \cap S \equiv \wedge) \qquad \text{Df} \qquad (3.77)$$

For example, the relation of *being a father of* excludes the relation of *being a brother to.*

(v) Worked example

The relation of inclusion among non-null classes excludes the relation of exclusion.

$$X \equiv \wedge \vee Y \equiv \wedge \vee X \not\equiv \wedge \qquad (3.78)$$

(1.33)

$$X \equiv \wedge \vee Y \equiv \wedge \vee X \cap (Y \cup Y') \not\equiv \wedge \qquad (3.79)$$

(1.34)

$$X \equiv \wedge \vee Y \equiv \wedge \vee X \cap Y' \not\equiv \wedge \vee X \cap Y \not\equiv \wedge \qquad (3.80)$$

(1.20)

$$(X \not\equiv \wedge \wedge Y \not\equiv \wedge)' \vee X \cap Y' \not\equiv \wedge \vee X \cap Y \not\equiv \wedge \qquad (3.81)$$

(1.11)

$$(X \not\equiv \wedge \wedge Y \not\equiv \wedge)' \vee (X \subset Y)' \vee X \cap Y \not\equiv \wedge \qquad (3.82)$$

(2.58) and (2.54)

$$(X \not\equiv \wedge \wedge Y \not\equiv \wedge)' \vee (X \subset Y \wedge X \cap Y \equiv \wedge)' \qquad (3.83)$$

(1.11)

$$(X \not\equiv \wedge \wedge Y \not\equiv \wedge) \Rightarrow (XKY \wedge XLY)' \qquad (3.84)$$

Definitions: (3.17) and (3.71)

$$(X \not\equiv \Lambda \,\Delta\, Y \not\equiv \Lambda) \Rightarrow K \cap L \equiv \Lambda \qquad (3.85)$$

Definitions: (2.50) and (2.67)

Note that in (3.85) the symbol $\Lambda$ stands for both the null-class of classes in general and the null-class of relations in particular.

---

**Exercises**

3.8 Show that the relation of exclusion among non-null classes is included in the relation of diversity.

3.9 Show that
(a) $(R \cap S)^{-1} \equiv R^{-1} \cap S^{-1}$
(b) $(R \cup S)^{-1} \equiv R^{-1} \cup S^{-1}$
(c) $R \subset S^{-1} \equiv R^{-1} \subset S$

3.10 Show that identity is included in the cartesian product of a non-null set with itself. Is the same true of diversity?

3.11 Show that if $R$ is non-null:
(a) $R \mid I \equiv R$
(b) $I \mid R \equiv R$

Why is it not true, in general, that $R \mid R^{-1} \equiv I$? Give instances both of the 'truth' and 'falsehood' of this equivalence.

3.12 Show that $R \mid (S \cup T) \equiv (R \mid S) \cup (R \mid T)$.
Is it the case that, in general, $R \mid (S \cap T) \equiv (R \mid S) \cap (R \mid T)$?

3.13 Show that $P \times (Q \cup R) \equiv (P \times Q) \cup (P \times R)$, i.e., that cartesian multiplication distributes over union.

3.14 Investigate the distribution of union and intersection over cartesian multiplication.

3.15 Show that $S \subset T \Rightarrow R \mid S \subset R \mid T$.

---

## 3.4 Sets of relations

Just as sets can be related (by inclusion, exclusion, etc.) when they are classified in ordered pairs, so also relations can be setted when they are classified singly. This section deals with one or two of the more important classifications; in accordance with convention three character alphabetic groups in lower case are used to signify sets of relations.

(a) *Reflexive, irreflexive and non-reflexive relations*

A relation is said to be **reflexive** if every member of its field bears this relation to itself:

$$R \in \text{rfl} \equiv (x)(x \in F(R) \Rightarrow xRx) \quad \text{Df} \quad (3.86)$$

For example, identify, the cartesian product of a set with itself, the relative product of a relation with its inverse, *being the same age as.*

A relation is said to be **irreflexive** if every member of its field bears the complementary relation to itself:

$$R \in \text{irr} \equiv (x)(x \in F(R) \Rightarrow xR'x) \quad \text{Df} \quad (3.87)$$

For example, diversity, *being older than.*

A relation is said to be **non-reflexive** if it is neither reflexive nor irreflexive:

$$R \in \text{nrf} \equiv (\exists x)((\exists y)(xRx \wedge y \in F(R) \wedge yR'y)) \quad \text{Df} \quad (3.88)$$

For example, *being the second cousin of;* a man may be his own second cousin or he may not.

(b) *Symmetric, asymmetric, antisymmetric and non-symmetric relations*

A relation is said to be **symmetric** if it is identical with its inverse:

$$R \in \text{sym} \equiv (R \equiv R^{-1}) \quad \text{Df} \quad (3.89)$$

For example, diversity, identity, overlapping, exclusion, *being a sibling of.*

A relation is said to be **asymmetric** if it is included in the complement of its inverse:

$$R \in \text{asm} \equiv R \subset (R^{-1})' \quad \text{Df} \quad (3.90)$$

For example, the null relation, the cartesian product of a set with its complement, *being older than* in a universe of individuals of diverse age.

Notice that we do not need $(R^{-1})' \subset R$.

A relation is said to be **antisymmetric** if the intersection of itself and its inverse implies (i.e., is included in) identity:

$$R \in \text{ant} \equiv R \cap R^{-1} \subset I \quad (3.91)$$

For example, inclusion.

A relation is said to be **non-symmetric** if it is neither symmetric nor asymmetric:

$$R \in \text{nsm} \equiv (R \cap R^{-1} \not\equiv \Lambda) \wedge (R \cap R^{-1} \not\equiv R) \quad \text{Df} \quad (3.92)$$

For example, *being a brother of.*

(i) Worked example

If a relation, $R$, is both symmetric and asymmetric it is null.

$$R^{-1} \equiv R \tag{3.93}$$
Premiss

$$R \subset (R^{-1})' \tag{3.94}$$
Premiss

$$R \subset R' \tag{3.95}$$
(3.93) and (3.94)

$$xRy \Rightarrow xR'y \tag{3.96}$$
Definition

$$(xRy \Rightarrow xR'y) \Rightarrow xR'y \tag{3.97}$$
(1.40)

$$xR'y \tag{3.98}$$
(3.96) and (3.97)

$$(x.y)(xR'y) \tag{3.99}$$
(3.98), (1.53) and (2.118)

$$R' \equiv \vee \tag{3.100}$$
(3.99) and (3.33)

$$R \equiv \wedge \tag{3.101}$$
(3.100) and (1.16)

(ii) Worked example

If a relation, $R$, is asymmetric then it is irreflexive. Take the field of $R$ as the universe of discourse.

$$(x)((y)(xRy \Rightarrow x(R^{-1})'y)) \tag{3.102}$$
Premiss

$$(x)((y)(xRy \Rightarrow yR'x)) \tag{3.103}$$
(3.31)

$$(x)(xRx \Rightarrow xR'x) \tag{3.104}$$
(2.116)

$$(x)(xR'x) \tag{3.105}$$
(1.40)

(iii) Worked example

If a relation $R$ is both irreflexive and transitive, it is asymmetric.

$$(w)(wR'w) \tag{3.106}$$
Premiss

$$(x,y,z)((xRy \wedge yRz) \Rightarrow xRz) \tag{3.107}$$
Premiss

$$aR'a \tag{3.108}$$
(3.106) and (2.116)

$$(aRb \wedge bRa) \Rightarrow aRa \tag{3.109}$$

(3.107) and (2.116)

$$(aRb \wedge bRa)' \tag{3.110}$$

(3.108), (3.109) and (1.41)

$$aRb \Rightarrow bR'a \tag{3.111}$$

(3.110) and (1.11)

$$(x,y)(xRy \Rightarrow yR'x) \tag{3.112}$$

(3.111), (1.53) and (2.118)

(c) *Transitive, intransitive and non-transitive relations*

A relation is said to be **transitive** if it includes the relative product of itself with itself:

$$R \in \text{trs} \equiv R \,|\, R \subset R \qquad \text{Df} \tag{3.113}$$

For example, inclusion, the cartesian product of a set with itself, *being a sibling of.*

A relation is said to be **intransitive** if its complement includes the relative product of itself with itself:

$$R \in \text{int} \equiv R \,|\, R \subset R' \qquad \text{Df} \tag{3.114}$$

For example, complementation, *being a parent of.*

A relation is said to be **non-transitive** if it is neither transitive nor intransitive:

$$R \in \text{ntr} \equiv ((R \,|\, R) \cap R \not\equiv \Lambda) \wedge (R \,|\, R) \cap R' \not\equiv \Lambda)) \qquad \text{Df} \tag{3.115}$$

For example, exclusion, non-equivalence, *being a cousin of.*

(d) *Equivalence relations and their domains*

Relations which are reflexive, symmetrical and transitive are known as **equivalence relations:**

$$\text{equ} \equiv \text{rfl} \cap \text{sym} \cap \text{trs} \qquad \text{Df} \tag{3.116}$$

For example, *living at the same address as.*

An equivalence relation, $R$, has the property that it subdivides the elements of its field into distinct sets, $E$, $F$, $G$, etc., such that every member of the field is a member of one and only one such set, and if $x$ and $y$ are two members of one such set then $xRy$.

To prove this denote by $E(x)$ the equivalence set generated under the equivalence relation, $R$, by any element, $x$, of its field as follows:

$$y \in E(x) \equiv xRy \qquad \text{Df} \tag{3.117}$$

We prove first that $x$ is a member of its own equivalence set. This follows from the fact that $R$ is reflexive. For, taking the field of $R$ as the universe of discourse:

$$(x)(xRx) \qquad (3.118)$$

Premiss

$$(x)(x \in E(x)) \qquad (3.119)$$

(3.118) and (3.117)

We now prove $(x)((y)(xRy \Rightarrow (E(x) \equiv E(y)))$

$$(xRy \wedge yRz) \Rightarrow xRz \qquad (3.121)$$

Premiss

$$(xRy \wedge z \in E(y)) \Rightarrow z \in E(x) \qquad (3.122)$$

Definition

$$xRy \Rightarrow (z \in E(y) \Rightarrow z \in E(x)) \qquad (3.123)$$

(3.122) and (1.44)

$$xRy \Rightarrow (z)(z \in E(y) \Rightarrow z \in E(x)) \qquad (3.124)$$

(3.123) and (2.118)

$$xRy \Rightarrow E(y) \subset E(x) \qquad (3.125)$$

(3.124) and Definition

$$xRy \equiv yRx \qquad (3.126)$$

Premiss

$$yRx \Rightarrow E(x) \subset E(y) \qquad (3.127)$$

(3.125: $y$ for $x$, $x$ for $y$)

$$xRy \Rightarrow E(x) \subset E(y) \qquad (3.128)$$

(3.126) and (3.127)

$$xRy \Rightarrow (E(x) \equiv E(y)) \qquad (3.129)$$

(3.125) and (3.128)

The rest of the proof as in (3.27)–(3.31).

The converse of the above follows:

$$(E(x) \equiv E(y)) \Rightarrow (z)(z \in E(x) \Rightarrow z \in E(y)) \qquad (3.130)$$

(2.162)

$$\Rightarrow (z)(xRz \Rightarrow yRz) \qquad (3.131)$$

Definition

$$yRz \equiv zRy \qquad (3.132)$$

Premiss

$$(E(x) = E(y)) \Rightarrow (z)(xRz \rightarrow zRy) \qquad (3.133)$$

(3.131) and (3.132)

$$\Rightarrow (z)(xRz \equiv (xRz \wedge zRy)) \qquad (3.134)$$

(3.133) and (1.37)

$$(E(x) \equiv E(y)) \Rightarrow (xRx \equiv (xRx \wedge xRy)) \qquad (3.135)$$

(3.134) and (2.116)

$$xRx \qquad (3.136)$$

Premiss

$$(E(x) \equiv E(y)) \Rightarrow (xRx \wedge xRy) \tag{3.137}$$

(3.135) and (3.136)

$$(E(x) \equiv E(y)) \Rightarrow xRy \tag{3.138}$$

(3.136) and (3.137)

It follows that two equivalence classes under a given equivalence relation, $R$, are either identical or have no common member:

Lemma $(z)(\phi z \Rightarrow \psi z) \Rightarrow ((\exists z)(\phi z) \Rightarrow (\quad z)(\psi z))$

$$(z)(\phi z \Rightarrow \psi z) \equiv (z)(\psi' z \Rightarrow \phi' z) \tag{3.140}$$

(1.15)

$$\Rightarrow ((z)(\psi' z) \Rightarrow (z)(\phi' z)) \tag{3.141}$$

(3.140) and (2.130)

$$\equiv ((\exists z)(\phi z) \Rightarrow (\exists z)(\psi z)) \tag{3.142}$$

(3.141), (1.15) and (2.5)

Theorem

$$(\exists z)(z \in E(x) \wedge z \in E(y)) \equiv (\exists z)(xRz \wedge yRz) \tag{3.143}$$

(3.117)

$$\equiv (\exists z)((E(x) \equiv (Ez)) \wedge (E(y)(E(y) \equiv E(z))) \tag{3.144}$$

(3.143), (3.121)–(3.129) and (3.130)–(3.138)

$$(z)(((E(x) \equiv E(z)) \wedge (E(y) \equiv E(z))) \Rightarrow (E(x) \equiv E(y))) \tag{3.145}$$

(2.    )

$$(\exists z)((E(x) \equiv E(z)) \wedge (E(y) \equiv E(z))) \Rightarrow (\exists z)(E(x) \equiv (E(y))$$

Lemma (3.146)

$$\Rightarrow (E(x) \equiv E(y)) \tag{3.147}$$

(2.42)

$$(\exists z)(z \in E(x) \wedge z \in E(y)) \Rightarrow (E(x) \equiv E(y)) \tag{3.148}$$

(3.143), (3.144), (3.146) and (3.147)

Finally, if $x$ and $y$ belong to the same equivalence class under a given equivalence relation, $R$, then $xRy$:

$$xRx \equiv xRz \tag{3.149}$$

Premiss

$$(xRz \wedge zRy) \Rightarrow xRy \tag{3.150}$$

Premiss

$$(x \in E(z) \wedge y \in E(z)) \equiv (zRx \wedge zRy) \tag{3.151}$$

Definition

$$\equiv xRz \wedge zRy \tag{3.152}$$

(3.149) and (3.151)

$$(x \in E(z) \wedge y \in E(z)) \Rightarrow xRy \tag{3.153}$$

(3.150), (3.151) and (3.152)

(e) *Connex and serial relations*

A relation is said to be **connex** if every pair of distinct members of its field are related either by the relation itself or its inverse:

$$R \in \text{cnx} \equiv J \subset R \cup R^{-1} \qquad \text{Df} \qquad (3.154)$$

For example, diversity, brotherhood in a family of siblings of which not more than one is a girl, *being greater than* in the field of natural numbers.

A relation is said to be **quasi-serial** if it is reflexive, antisymmetric and transitive:

$$\text{qsr} \equiv \text{rfl} \cap \text{ant} \cap \text{trs} \qquad \text{Df} \qquad (3.155)$$

For example, inclusion.

A relation is said to be **serial** if it is anti-symmetrical, transitive and connex:

$$\text{ser} \equiv \text{ant} \cap \text{trs} \cap \text{cnx} \qquad \text{Df} \qquad (3.156)$$

For example, *being greater than or equal to* in the field of natural numbers.

A relation is said to be **strictly serial** if it is asymmetrical, transitive and connex:

$$\text{ssr} \equiv \text{asm} \cap \text{trs} \cap \text{cnx} \qquad \text{Df} \qquad (3.157)$$

For example, *being greater than* in the field of natural numbers.

---

**Exercises**

3.16 Show that all intransitive relations are irreflexive.

3.17 Take $F$ to stand for fatherhood, $H$ for *being the husband of* and $S$ for sonship. Determine in terms of these relations, their intersections, unions, inverses, complements and relative products, suitable expressions for:

(a) daughterhood
(b) motherhood
(c) sisterhood
(d) brotherhood
(e) first-cousinship

in the universe of human beings, alive or dead. Which of these relations, if any, are reflexive, transitive, symmetrical?

3.18 Show that if a relation is serial so is its inverse. Is the same true if the relation is quasi-serial?

3.19 Show that $R \in \text{asm} \equiv R|R \in \text{irr}$

3.20 If $R$, $S$ are equivalence relations, show that $R \cap S$ is also.

3.21 Show that if $R$ is reflexive and transitive and if $a, b \in S \equiv (aRb \wedge bRa)$, then $S$ is an equivalence relation.

---

## 3.5 Relations between sets whose members are related

### (a) *Limited domains*

In section **3.2**(g) we defined $D(R)$, $C(R)$ and $F(R)$ as the sets whose members bore the relation $R$ to anything, bore the relation $R^{-1}$ to anything and bore the relation $R \cup R^{-1}$ to anything. It is now convenient to extend the definitions to the case where the 'anything' is restricted to a specific set, $S$.

$$x \in D(R, S) \equiv (\exists y)(y \in S \wedge xRy) \quad \text{Df} \quad (3.158)$$
$$x \in C(R, S) \equiv (\exists y)(y \in S \wedge yRx) \quad \text{Df} \quad (3.159)$$
$$x \in F(R, S) \equiv (\exists y)(y \in S \wedge (xRy \vee yRx) \quad \text{Df} \quad (3.160)$$

It should be noted, though it is obvious from their form, that in these definitions $R$ stands for a relation, $S$ for a set.

Worked example

$$x \in D(R, D(R)) \equiv x \in D(R|R) \quad (3.161)$$

$$x \in D(R, D(R)) \equiv (\exists y)(y \in D(R) \wedge xRy) \quad (3.162)$$
(3.158)
$$\equiv (\exists y)((\exists z)(yRz) \wedge xRy) \quad (3.163)$$
(3.162) and (3.66)
$$\equiv (\exists y)((\exists z)(yRz \wedge xRy)) \quad (3.164)$$
(3.163) and (2.42)
$$\equiv (\exists z)((\exists y)(yRz \wedge yRz)) \quad (3.165)$$
(3.164) and (3.54)
$$\equiv (\exists z)((\exists y)(xRy \wedge yRz)) \quad (3.166)$$
(3.165) and (1.8)
$$\equiv (\exists z)(x(R|R)z) \quad (3.167)$$
(3.166) and (3.36)
$$\equiv x \in D(R|R) \quad (3.161)$$
(3.167) and (3.66)

$$x \in D(R, D(R, S)) \equiv x \in D(R|R,S) \quad (3.168)$$

This follows by an obvious extension of the proof of (3.161).

(b) *Fundamental theorems of limited domains*

$$S \subset T \Rightarrow D(R,S) \subset D(R,T)$$

$$S \subset T \equiv (x)(x \in S \Rightarrow x \in T) \qquad (3.171)$$

(2.58)

$$\Rightarrow (x)((x \in S \wedge wRx) \Rightarrow (x \in T \wedge wRx)) \qquad (3.172)$$

(3.171) and (1.25)

$$\Rightarrow ((\exists y)(y \in S \wedge wRy) \Rightarrow (\exists z)(z \in T \wedge wRz)) \qquad (3.173)$$

(3.172) and (3.142)

$$\Rightarrow (w)((\exists y))(y \in S \wedge wRy) \Rightarrow (\exists z)(z \in T \wedge wRz)) \qquad (3.174)$$

(3.173) and (2.118)

$$\Rightarrow (w)(w \in D(R,S) \Rightarrow w \in D(R,T)) \qquad (3.175)$$

(3.174) and (3.158)

$$\Rightarrow D(R,S) \subset D(R,T) \qquad (3.176)$$

(3.175) and (2.58)

For example, if $R$ stands for *is the head of*, $S$ for the set of horses and $T$ for the set of animals, then *all horses are animals* implies *the head of an horse is the head of an animal.*

$$S \subset T \Rightarrow C(R,S) \subset C(R,T) \qquad (3.177)$$

This follows from (3.176) by substituting $R^{-1}$ for $R$.

$$S \subset T \Rightarrow F(R,S) \subset F(R,T) \qquad (3.178)$$

This follows from (3.176) by substituting $R \cup R^{-1}$ for $R$.

(c) *Uniquely limited domains*

A special case of (3.158) occurs when $S$ is a unit set, say $[a]$. In this case we prove:

$$x \in D(R, [a]) \equiv xRa$$

We need a lemma, viz:

$$((x)(\phi x \Rightarrow p) \wedge (\exists x)(\phi x)) \Rightarrow p \qquad (3.180)$$

Lemma

$$((x)(\phi x \Rightarrow p) \wedge (\exists x)(\phi x)) \Rightarrow (\exists x)(\phi x \wedge (\phi x \Rightarrow p)) \qquad (3.181)$$

(2.37)

$$\Rightarrow (\exists x)(\phi x \wedge p) \qquad (3.182)$$

(3.181) and (1.42)

$$\Rightarrow (\exists x)(\phi x) \wedge p \tag{3.183}$$
(3.182) and (2.42)
$$\Rightarrow p \tag{3.184}$$
(3.183) and (1.6)

Theorem

$$((y)((y = a \wedge xRy) \Rightarrow xRa) \wedge (\exists y)(y = a \wedge xRy)) \Rightarrow xRa \tag{3.185}$$
(3.180)
$$(y)(y = a \wedge xRy) \Rightarrow (y = a \wedge xRy) \tag{3.186}$$
(2.116)
$$\Rightarrow xRa \tag{3.187}$$
(3.186) and (3.12)
$$(\exists y)(y = a \wedge xRy) \Rightarrow xRa \tag{3.188}$$
(3.185) and (3.187)
$$(\exists y)(y \in [a] \wedge xRy) \Rightarrow xRa \tag{3.189}$$
(3.188) and (2.212)–(2.218)
$$x \in D(R, [a]) \Rightarrow xRa \tag{3.190}$$
(3.189) and (3.158)
$$xRa \equiv (a \in [a] \wedge xRa) \tag{3.191}$$
(1.34)
$$\Rightarrow (\exists y)(y \in [a] \wedge xRy) \tag{3.192}$$
(3.191) and (2.41)
$$\Rightarrow x \in D(R,[a]) \tag{3.193}$$
(3.192) and (3.158)
$$xRa \equiv x \in D(R, [a]) \tag{3.194}$$
(3.190) and (3.193)
$$x \in C(R,[a]) \equiv aRx \tag{3.195}$$
(3.194: $R^{-1}$ for $R$)
$$x \in F(R,[a]) \equiv xRa \vee aRx \tag{3.196}$$
(3.194: $R \cup R^{-1}$ for $R$)

(d) *Functions*

A relation $R$ is said to be **one–many**, or to constitute a **many–valued function** of its domain on to its counter-domain if:

$$R \mid R^{-1} \subset I \tag{3.197}$$

This type of relation is illustrated in Figure 3.1. Examples are the cartesian product of a unit set with the universal set, and fatherhood.

A relation $R$ is said to be **many–one** if its inverse, $R^{-1}$, is one–many. Finally, a relation $R$ is said to be **one–one** or **biunique** if:

$$(R \mid R^{-1}) \cup (R^{-1} \mid R) \subset I \tag{3.198}$$

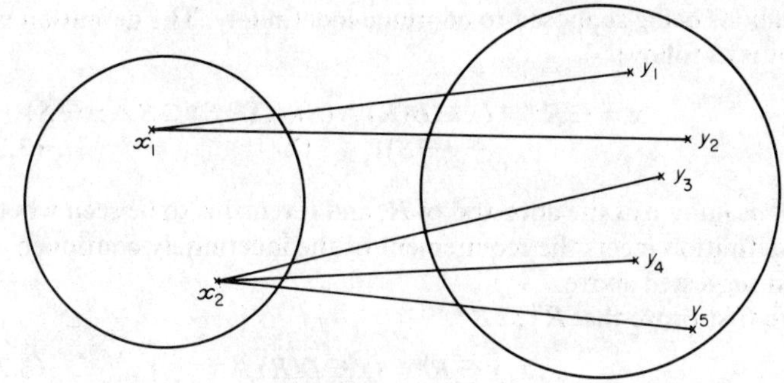

Fig. 3.1

This is incorporated in the following definition:

$$R \in \text{bnq} \equiv (R \mid R^{-1} \subset I) \wedge (R^{-1} \mid R \subset I) \quad \text{Df} \qquad (3.199)$$

Such a relation constitutes a **single-valued** or **unique function** of both its domain and its counter-domain, each on the other. This is illustrated in Figure 3.2. For example, complementation, inversion, *being the spouse of* (in a monogamous universe).

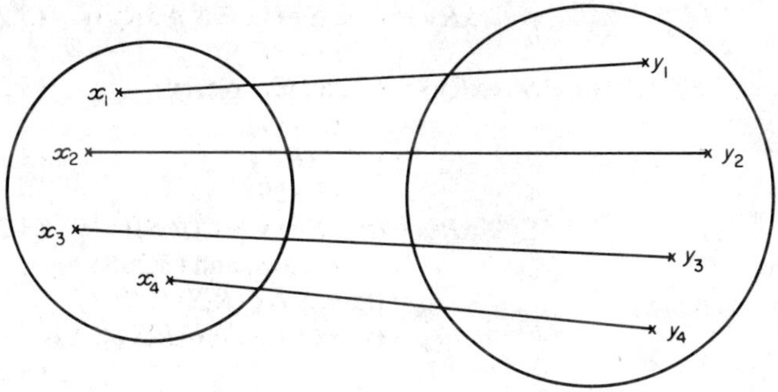

Fig. 3.2

(e) *The ancestral relation of R*

Suppose we denote by $R^0$ the relation of identity in the field of $R$; by $R^2$ the relative product of $R$ with itself; by $R^3$ the relative product of $R^2$ with $R$ (see Note 8); and so on. We now wish to define a relation which shall include just $R^0$, $R$, $R^2$, $R^3$, and so on, i.e., we wish to obtain a circumscribed definition of R* so that:

$$R^* \equiv R^0 \cup R \cup R^2 \cup R^3 \cup \ldots \qquad (3.200)$$

the unions being supposed to continue indefinitely. The definition we adopt is as follows:

$$x, y \in R^* \equiv (x \in D(R) \wedge (S)((C(R,S) \subset S \wedge x \in S) \Rightarrow y \in S)) \quad \text{Df} \qquad (3.201)$$

$R^*$ is known as the **ancestral** of $R$, and it remains to be seen whether this definition meets the requirement of the indefinitely continued union suggested above.

We first prove that $R^0 \subset R^*$.

$$x, y \in R^0 \equiv (x \in D(R) \wedge x \equiv y) \qquad (3.202)$$

Definition

$$\Rightarrow (x \in D(R) \wedge (S)(x \in S \Rightarrow y \in S)) \qquad (3.203)$$

(3.202) and (1.25)

$$\Rightarrow (x \in D(R) \wedge (S)((C(R,S) \subset S \wedge x \in S) \Rightarrow y \in S))) \qquad (3.204)$$

(3.203) and (1.6)

$$R^0 \subset R^* \qquad (3.205)$$

(3.204) and (3.201)

We now show that $R \subset R^*$

$$xRy \Rightarrow (x \in S \Rightarrow (x \in S \wedge xRy)) \qquad (3.206)$$

(1.43)

$$(x \in S \wedge xRy) \Rightarrow (\exists z)(z \in S \wedge zRy) \qquad (3.207)$$

(2.117)

$$\Rightarrow y \in C(R,S) \qquad (3.208)$$

(3.159)

$$xRy \Rightarrow (x \in S \Rightarrow y \in C(R,S)) \qquad (3.209)$$

(3.206) and (3.208)

$$\Rightarrow ((x \in S \wedge C(R,S) \subset S) \Rightarrow (y \in C(R,S) \wedge C(R,S) \subset S)) \qquad (3.210)$$

(3.209) and (1.25)

$$(y \in C(R,S) \wedge C(R,S) \subset S) \Rightarrow y \in S \qquad (3.211)$$

(2.58) and (2.116)

$$xRy \Rightarrow ((x \in S \wedge C(R,S) \subset S) \Rightarrow y \in S) \qquad (3.212)$$

(3.210) and (3.211)

$$xRy \Rightarrow (S)((x \in S \wedge C(R,S) \subset S) \Rightarrow y \in S) \qquad (3.213)$$

(3.212) and (2.118)

$$xRy \Rightarrow x \in D(R) \qquad (3.214)$$

(3.66) and (2.117)

$$xRy \Rightarrow (x \in D(R) \wedge (S)((x \in S \wedge C(R,S) \subset S) \Rightarrow y \in S) \qquad (3.215)$$

(3.213) and (3.214)

$$R \subset R^* \qquad (3.216)$$

(3.215) and (3.201)

To extend the argument of the preceding paragraph so as to show that $R^2 \subset R^*$ we proceed as follows:

$$C(R,S) \subset S \Rightarrow C(R,C(R,S)) \subset C(R,S) \qquad (3.217)$$

(3.177: $C(R,S)$ for $S$, $S$ for $T$)

$$C(R,C(R,S)) \equiv C(R^2,S) \qquad (3.218)$$

(3.169: $R^{-1}$ for $R$)

$$C(R,S) \subset S \Rightarrow C(R^2,S) \subset C(R,S) \qquad (3.219)$$

(3.217) and (3.218)

$$(x \in S \wedge C(R,S) \subset S) \Rightarrow (x \in S \wedge C(R^2,S) \subset C(R,S)) \qquad (3.220)$$

(3.219) and (1.25)

$$xR^2y \Rightarrow ((x \in S \wedge C(R^2,S) \subset S) \Rightarrow y \in S) \qquad (3.221)$$

(3.212: $R^2$ for $R$)

$$xR^2y \Rightarrow ((x \in S \wedge C(R,S) \subset S) \Rightarrow y \in S) \qquad (3.222)$$

(3.220) and (3.221)

$$xR^2y \Rightarrow (S)((x \in S \wedge C(R,S) \subset S) \Rightarrow y \in S) \qquad (3.223)$$

(3.222) and (2.118)

$$xR^2y \Rightarrow (\exists z)(xRz \wedge zRy) \qquad (3.224)$$

(3.36)

$$\Rightarrow ((\exists z)(xRz) \wedge (\exists z)(zRy)) \qquad (3.225)$$

(2.25)

$$\Rightarrow (\exists z)(xRz) \qquad (3.226)$$

(1.104)

$$\Rightarrow x \in D(R) \qquad (3.227)$$

(3.66)

$$xR^2y \Rightarrow x \in D(R) \wedge (S)((x \in S \wedge C(R,S) \subset S) \Rightarrow y \in S)) \qquad (3.228)$$

(3.223) and (3.227)

$$R^2 \subset R^* \qquad (3.229)$$

(3.228) and (3.201)

The substance of this argument is illustrated in Figure 3.3. It is plain that the argument can be similarly extended to show that $R^3 \subset R^*$, $R^4 \subset R^*$, etc. But what exactly is this 'etc'? To formulate the general result we should have to say that if $n$ was a natural number (positive

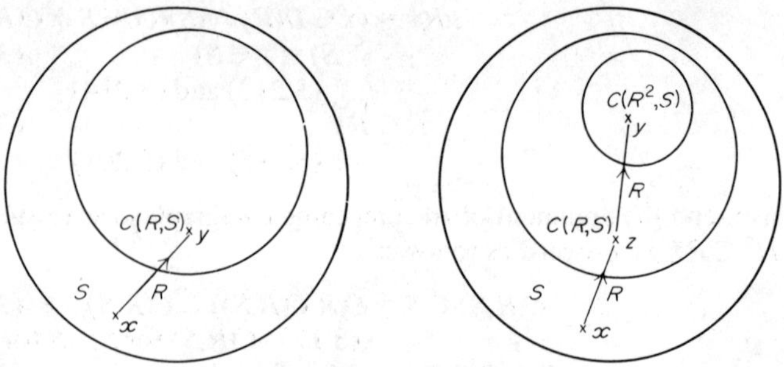


Fig. 3.3

integer or zero) then $R^n \subset R^*$. But we have no definition of a natural number, and it is precisely to the ancestral relation that we turn to obtain one. In doing so we are able to show not merely that if $x,y$ is a member of $R^*$, then there exists a natural number, $n$, such that $x,y \in R^n$ (that now becomes a matter of definition), but also, and conversely, that for any natural number $n$, $R^n \subset R^*$. To this, the last task we have set ourselves, we now turn.

---

**Exercises**

3.22 Show that $(R^{-1})^* \equiv (R^*)^{-1}$.

3.23 Show that $R^0 \cup (R \mid R^*) \equiv R^*$. How can you extend this?

3.24 Show that $(R^*)^* \equiv R^*$. What other relations are their own ancestrals?

3.25 Show by counter-examples that $(R \cup S)^* \not\equiv R^* \cup S$ and $(R \cap S)^* \not\equiv R^* \cap S^*$, for some $R,S$.

3.26 Show that $(R|S)^* \subset R^*|S^*$.

3.27 Show that if $R$, $S$ are biunique and their domains and counter-domains mutually exclusive, $R \cup S$ is biunique also.

3.28 Show that if $R$ is biunique and $S \subset R$, $S$ is also.

---

**Notes on chapter 3**

1. In this chapter we have taken the notion of an ordered pair – $x,y$ – as primitive, i.e., beyond the scope of definition, since this notion or something very like it was needed in the analysis of a

primitive proposition in section 2.2(a). However, in many works on logic the ordered pair is defined as the class whose sole members are $[x]$ and $[x] \cup [y]$, i.e.:

$$\langle x,y\rangle \equiv [x] \cup [[x] \cup [y]] \quad \text{Df}$$

From this definition it is not difficult to infer the equivalence of (3.12) (there an equivalence of definition), namely:

$$\langle x,y\rangle \equiv \langle u,v\rangle \equiv (x = u \wedge y = v)$$

provided care is taken to distinguish the cases $x = y$, $x \neq y$. But even when the ordered pair is made the subject of definition in this way we still need something like (3.3) to justify the use of $\langle x,y\rangle$ as a single variable of quantification.

2. 'Relationship' is here used in preference to 'relation' since in common speech the latter often refers to the individuals related. From section 3.2(b) onwards, however, the term 'relation' is used exclusively.

3. The cases of three or more variables may be reduced to the case of two by defining the ordered triad by means of:

$$x,y,z \equiv (x,y),z$$

the ordered tetrad by means of:

$$x,y,z,w \equiv ((x,y),z),w$$

and so on.

4. German capitals are often used to denote relations. But this usage tends to obscure the fact that a relation is in fact a class of ordered pairs.

5. Sometimes the complementary relation $R'$ is termed the **converse** of $R$.

6. *Hereditary* is here used in the sense of *disposing of the kingdom by inheritance.*

7. Notice that we cannot make this our definition of an ordered pair $x,y$ since we need that notion in order to define the cartesian product of two sets.

8. This definition of $R^3$ is unambiguous since $R^2|R \equiv R|R^2$ by (3.65)

**Definitions used in chapter 3**

(27) $(x,y)(\phi x,y)$ for $(x)((y)(\phi x,y))$
(28) $x,y = u,v$ for $x = u \wedge y = v$
(29) $xRy$ for $x,y \in R$
(30) $x,y \in I$ for $(\exists R)(xRy \wedge (i,j)(iRj \Rightarrow i = j))$
(31) $x,y \in J$ for $x,y \notin I$

(32) $x,y \in R'$ for $x,y \notin R$
(33) $x,y \in R^{-1}$ for $y,x \notin R$
(34) $x,y \in P \times Q$ for $x \in P \wedge y \in Q$
(35) $x,y \in R \mid S$ for $(\exists z)(xRz \wedge zSy)$
(36) $x \in D(R)$ for $(\exists y)(xRy)$
(37) $x \in C(R)$ for $(\exists y)(yRx)$
(38) $x \in F(R)$ for $x \in C(R) \cup D(R)$
(39) $R \in$ rfl for $(x)(x \in F(R) \Rightarrow xRx)$
(40) $R \in$ irr for $(x)(x \in F(R) \Rightarrow xR'x)$
(41) $R \in$ nrf for $(\exists x)((\exists y)(xRx \wedge y \in F(R) \wedge yR'y))$
(42) $R \in$ sym for $R \equiv R^{-1}$
(43) $R \in$ asm for $(R \subset R^{-1})'$
(44) $R \in$ ant for $R \cap R^{-1} \subset I$
(45) $R \in$ nsm for $(R \cap R^{-1} \not\equiv \Lambda) \wedge (R \cap R^{-1} \not\equiv R)$
(46) $R \in$ trs for $R \mid R \subset R$
(47) $R \in$ int for $R \mid R \subset R'$
(48) $R \in$ ntr for $(R \mid R) \cap R \not\equiv \Lambda \wedge (R \mid R) \cap R' \wedge \Lambda$
(49) equ for rfl $\cap$ sym $\cap$ trs
(50) $R \in$ cnx for $J \subset R \subset R^{-1}$
(51) qsr for rfl $\cap$ ant $\cap$ trs
(52) ser for ant $\cap$ trs $\cap$ cnx
(53) ssr for asm $\cap$ trs $\cap$ cnx
(54) $x \in D(R,S)$ for $(\exists y)(y \in s \wedge xRy)$
(55) $x \in C(R,S)$ for $(\exists y)(y \in S \wedge yRx)$
(56) $x \in F(R,S)$ for $(\exists y)(y \in S \wedge (xRy \vee yRx))$
(57) $R \in$ bnq for $(R \mid R^{-1} \subset I) \wedge (R^{-1} \mid R \subset I)$
(58) $x,y \in R^*$ for $x \in D(R) \wedge (S)((C(R,S) \subset S \wedge x \in S) \Rightarrow y \in S)$

# 4

# THE LOGIC OF ARITHMETIC

## 4.1 Counting and measuring

The few instances of scientific 'law' which have been cited in previous chapters – the law of gravity, Boyle's law and Mendel's law of inheritance – all depend for their verification either upon counting or measuring or both. Since it is natural to regard measurement as an extension of counting to the case where subdivisions of the units counted – whether of length, time or mass – are counted in their turn, we can fairly say that the fundamental activity in all scientific investigation is counting.

But what is counting? It is the object of this chapter to show that it is essentially a logical operation; that the concepts involved in it are no different from those we have so far encountered – proposition, set and relation; and that all the so-called rules of arithmetic are in fact the logical consequences of a suitable definition of **number**. For the purposes of our argument we take 'number' to mean 'natural number', i.e., 0, 1, 2, etc., and for the sake of lucidity the argument itself is presented less formally, though not less rigorously than hitherto.

### (a) *Learning to count*

When a child is taught to count, two separate techniques are employed and it is important to distinguish them. The first is the recognition of the similarity between groups having the same number of objects in them – five hens, five ducks. The second (though often taught first) is the listing of the numbers 'one', 'two', 'three' and so on up to a given point in the correct order.

A child could of course be taught the meaning of 'number' by means of the first technique alone. Some people can recognise quite large numbers of objects – up to thirty or forty in extreme instances – without counting them off at all. it is clear therefore that we must distinguish between numbers as such and numbers placed in some order. It is so natural to think of the numbers 'one', 'two', 'three', in that order that it is at first hard to think of them in any other order or in no order at all. Nevertheless if we want a definition of number as such we must for the moment leave out of account the question of the order (if any) in which particular numbers may be arranged.

To do this let us fix our attention on the relation between a group of five hens and a group of five ducks. What is common to these groups? Simply the property of being five in number. And this is a property of the groups – not of their individual members. We may say of the Ten Commandments that each Commandment is Mosaic, but not that each is ten. Thus number appears as a property[1] of groups or, in the language of logic, as a class of classes. The number 'five' will therefore be defined as **the class of all classes that are equinumerate with some class having five members**.

At first sight this definition appears circular, as though it defined number in terms of itself. But is it not so; for two reasons. First we can define **equinumeracy** (or **similarity** as it is more often termed) without reference to any previous notion of number. Indeed we can make equinumeracy the basis of a definition of number itself. And to arrive at the definition of any particular number we can start with 'zero' defined as the class of all classes having no members and, by a suitable sequence, arrive at definitions of 'one' (the class of all unit classes), 'two' (the class of all pairs), 'three', 'four', 'five', and so on. This is in fact what the child is taught to do – beginning at one – and it is throughout an entirely logical procedure.

(b) *The relation of similarity*

When a child recognises that a group of five hens has something in common with a group of five ducks, it does so by mentally pairing off each hen with a duck in such a way that no hen is without a corresponding duck and vice versa. How this is done or in how many ways it might be done (for example by relating size of hen to size of duck or their relative positions on a page) is immaterial. All that matters is that it can be done, i.e., that there is at least one biunique relation having the set of hens as its domain and the set of ducks as its counter-domain. In general any two classes $P$, $Q$ are said to be similar or equinumerate if there exists a biunique relation $R$, having $P$ as its domain and $Q$ as its counter-domain.

$$P \text{ sim } Q \underset{\text{Df}}{\equiv} (\exists R)((P \equiv D(R)) \wedge (Q \equiv C(R)) \wedge R \in \text{bnq}) \tag{4.1}$$

Note that in this definition the alphabetic triad 'sim' stands for a relation of classes while the triad 'bnq' stands for a class of relations.

By taking $I$ for $R$ we may infer:

$$P \text{ sim } P \vee P \equiv \Lambda \tag{4.2}$$

By taking $R^{-1}$ for $R$ we may similarly infer:

$$P \text{ sim } Q \equiv Q \text{ sim } P \tag{4.3}$$

We can furthermore make use of (4.1) to define membership in the number, $N(P)$, of a class $P$ as follows:

$$Q \in N(P) \equiv P \text{ sim } Q \quad \text{Df} \tag{4.4}$$

Like every other class the number of a class is defined only in use. Evidently:

$$P \in N(P) \vee P \equiv \Lambda \tag{4.5}$$

(4.2) and (4.4)

and:

$$Q \in N(P) \equiv P \in N(Q) \tag{4.6}$$

(4.3) and (4.4)

We can also define number in general by means of:

$$x \in \text{num} \equiv (\exists P)(x = N(P)) \quad \text{Df} \tag{4.7}$$

while the identity of two numbers is expressed by:

$$N(P) = N(Q) \equiv P \text{ sim } Q \tag{4.8}$$

(4.4) and (2.162)

(c) *Similarity and its equivalence classes*

As might be expected similarity turns out to be an equivalence relation (section **3.4**(d)).

$$\text{sim} \in \text{rfl} \tag{4.9}$$

(3.86) and (4.2)

$$\text{sim} \in \text{sym} \tag{4.10}$$

(3.89) and (4.3)

$$\text{sim} \in \text{trs} \tag{4.11}$$

To prove this we require a lemma, namely:

$$((R \in \text{bnq}) \wedge (S \in \text{bnq})) \Rightarrow R \mid S \in \text{bnq} \tag{4.12}$$

$$S \in \text{bnq} \Rightarrow S \mid S^{-1} \subset I \tag{4.13}$$

(3.199)

$$S \in \text{bnq} \Rightarrow R \mid S \mid S^{-1} \subset R \mid I \tag{4.14}$$

(4.13) and (Exercise 3.15)

$$R \mid I \subset R \tag{4.15}$$

(Exercise 3.11)

$$S \in \text{bnq} \Rightarrow R \mid S \mid S^{-1} \subset R \tag{4.16}$$

(4.14) and (4.15)

$$S \in \text{bnq} \Rightarrow R \mid S \mid S^{-1} \mid R^{-1} \subset R \mid R^{-1} \tag{4.17}$$

(4.16) and (Exercise 3.15)

$$R \in \text{bnq} \equiv R \mid R^{-1} \subset I \tag{4.18}$$

(3.199)

$$(R \in \text{bnq} \wedge S \in \text{bnq}) \Rightarrow R|S|S^{-1} \subset I \tag{4.19}$$

(4.17) and (4.18)

$$\Rightarrow (R|S)|(R|S)^{-1} \subset I \tag{4.20}$$

(3.37)

$$(R \in \text{bnq} \wedge S \in \text{bnq}) \Rightarrow (R|S)^{-1}|(R|S) \subset I \tag{4.21}$$

(similarly)

$$(R \in \text{bnq} \wedge S \in \text{bnq}) \Rightarrow R|S \in \text{bnq} \tag{4.22}$$

(4.20), (4.21) and (3.199)

The theorem is proved as follows:

$$P \text{ sim}|\text{sim } Q \equiv (\exists T)(P \text{ sim } T \wedge T \text{ sim } Q) \tag{4.23}$$

(3.36)

$$\equiv (\exists T)((\exists R)((\exists S)(P \equiv D(R) \wedge T \equiv C(R) \wedge T \equiv D(S) \wedge Q \equiv C(S) \wedge R \in \text{bnq} \wedge S \in \text{bnq}))) \tag{4.24}$$

(4.23) and (4.1)

$$(\exists R)((\exists S)(P \equiv D(R) \wedge T \equiv C(R) \wedge T \equiv D(S) \wedge Q \equiv C(S))) \Rightarrow (\exists R)((\exists S)(P \equiv D(R) \wedge Q \equiv C(S))) \tag{4.25}$$

(1.104)

$$(\exists T)((\exists R)((\exists S)(P \equiv D(R) \wedge T \equiv C(R) \wedge T \equiv D(S) \wedge Q \equiv C(S)))) \Rightarrow (\exists R)((\exists S)(P \equiv D(R) \wedge Q \equiv C(S)) \tag{4.26}$$

(4.25) and (2.118)

$$P \text{ sim}|\text{sim } Q \Rightarrow (\exists R)((\exists S)(P \equiv D(R) \wedge Q \equiv C(S) \wedge R \in \text{bnq} \wedge S \in \text{bnq}) \tag{4.27}$$

(4.24) and (4.26)

$$U \equiv R|S \tag{4.28}$$

Definition

$$D(R) \equiv D(U) \tag{4.29}$$

(4.28), (3.36) and (3.66)

$$C(S) \equiv C(U) \tag{4.30}$$

(4.28), (3.36) and (3.67)

$$P \text{ sim}|\text{sim } Q \Rightarrow (\exists U)(P \equiv D(U) \wedge Q \equiv C(U) \wedge U \in \text{bnq}) \tag{4.31}$$

(4.27)–(4.30) and (4.12)

$$\Rightarrow P \text{ sim } Q \tag{4.32}$$

(4.31) and (4.1)

$$\text{sim}|\text{sim} \subset \text{sim} \tag{4.33}$$

(4.32) and (2.58)

(4.9)–(4.11) show that similarity is an equivalence relation (3.116); its equivalence classes are of course just the numbers, $N(P)$, defined by (4.4).

(d) *Number and natural number*

The relation of equinumeracy which has been used to define number in general does not confine that definition to finite numbers, i.e., such as we can reach by counting from zero. Nor does it place the numbers which it defines in any order. The radii of a circle can be placed in a biunique relation with the points on its circumference and these in turn with the points on the circumference of a concentric circle. But this does not mean that it is possible to count the numbers in either group or to compare their number with, say, the number of points within the circle. In dealing with the numbers that can be reached by counting, that is the **natural numbers**, we are dealing with only a subset of the numbers in general.

## 4.2 The relation of succession

For the rest of this chapter we are to confine our attention to the natural numbers.[3] To define these we fix on one particular number, zero, and then define its successors which we name 'one', 'two', etc.

(a) *The number zero*

It may seem unnatural to start with zero rather than one, more particularly as we have to extend our definition of number so as to include it. But it is in fact convenient to regard zero as a number. It is defined as the class of classes whose sole member is the null class:

$$x \in 0 \equiv x = \Lambda \qquad \text{Df} \tag{4.34}$$

Notice that since $\Lambda$ is a class this definition does not identify $\Lambda$ with 0; it does of course identify 0 with $[\Lambda]$ (2.219)

(b) *Immediate succession*

To define the successor of a natural number, i.e., the number which follows it in the natural order, we merely have to replace similarity to a given class by similarity to a class having one additional member. This means to say that if $z = N(P)$, the successor of $z$ ($y$, say) will be the number of any class having as its members all the members of $P$ and one other. In other words if $y$ is the successor of $z$, $y$ will be the class of all classes $x$ such that some element $w$ belongs to $x$ whose removal will yield a class that belongs to $z$. Using $S$ to symbolise the relation of succession:[4]

$$ySz \underset{\text{Df}}{=} (x)(x \in y \equiv (\exists w)(w \in x \wedge x \cap \lfloor w \rfloor' \in z)) \tag{4.35}$$

Four points should be noted in regard to this definition. First, it is no part of the definition that $z$ itself is a number; it may be any class of classes. Second, it can be shown (4.86)–(4.113) that if $z$ is a number then its successor is also a number and a unique number (provided the necessary element $w$ exists), since all the classes $x$ which satisfy the definition are similar to each other. It can also be shown (4.121(–(4.14) that no two natural numbers have the same successor. Finally, there is no number of which zero is the successor (4.141)–(4.149),

Proofs of these assertions are deferred until the next section so as not to hold up the course of the argument as a whole.

We are now in a position to define 1 as the successor of 0, 2 as the successor of 1 and so on as far as we like.

$$x \in 1 \equiv (\exists w)(w \in x \wedge x \cap [w]' \in 0) \quad \text{Df} \quad (4.36)$$
$$x \in 2 \equiv (\exists w)(w \in x \wedge x \cap [w]' \in 1) \quad \text{Df} \quad (4.37)$$
$$x \in 3 \equiv (\exists w)(w \in x \wedge x \cap [w]' \in 2) \quad \text{Df} \quad (4.38)$$

(c) *Natural numbers and their properties*

The definition of the succcessor of $z$ (4.34) is of course a definition of immediate succession. In dealing with the natural numbers however we are more interested in succession in a wider sense – in the sense, that is, that any number greater than or equal to another number may be said to be its successor. We want a relation to the first number, in other words, that shall be borne not only by the successor of that number as defined above but by the successor of the successor, the successor of that and so on. Such a relation is $(S^{-1})^*$, the ancestral of $S^{-1}$. If $xS^{-1}y$, that is $ySx$, we refer to $x$ as the **predecessor** of $y$ and write.[5]

$$xPy \equiv ySx \qquad \text{Df} \qquad (4.39)$$

If, further, $x(S^{-1})^*y$ or $xP^*y$ we say that $y$ belongs to the **posterity** of $x$ with respect to the relation $S$.

$$x \leqslant y \equiv xP^*y \qquad \text{Df} \qquad (4.40)$$

The relation $P^*$ – more often written as $\leqslant$, and pronounced *less than or equal to* – advances our concept of natural number in two ways. First it can be used to define the complete set of natural numbers. So far we have succeeded in defining number itself, but without limiting the definition to natural numbers; and we have defined the particular numbers 0, 1, 2, 3. . . .. The posterity of 0 with respect to $S$ provides a definition of natural number which is complete in itself.

Before adopting this definition, however, we must show that any particular number we may care to define by means of (4.36) et seq. will in fact belong to the set of natural numbers so defined. This amounts to a generalisation of the results obtained in (3.217)–(3.229) with $P(\equiv S^{-1})$ replacing $R$.

First, we suppose the domain of $S$ to be confined to the class of numbers, that is we take the definition of $S$ (4.35) to include the assertion that either $z$ is zero or it is a class of similar classes. This has the effect of identifying the field of $S$ with that of $S^{-1}$.

Next we define the **hereditary properties** (or hereditary classes[1]) with respect to a relation $R$ as follows:

$$Q \in \text{her}(R) \equiv C(R,Q) \subset Q \qquad (4.41)$$

Definition

$$\equiv (x)(x \in C(R,Q) \Rightarrow x \in Q) \qquad (4.42)$$

(2.58)

$$\equiv (x)((y)((y \in Q \wedge yRx) \Rightarrow x \in Q) \qquad (4.43)$$

(3.67)

This means that a property which is hereditary with respect to the relation *predecessor* is such that if it belongs to any number it belongs to the successor of that number.

If we now substitute, $P$, i.e., the relation *predecessor*, for $R$ in the definition of the ancestral (3.201) we observe that the members of the counter-domain of $P^*$ – that is the posterity of some $x$ – are numbers having just these hereditary properties. For, writing $P$ for $R$ and $Q$ for $S$ in that definition we have:

$$xPy^* \equiv (x \in D(P) \wedge (Q)((Q \in \text{her }(P) \wedge x \in Q) \Rightarrow y \in Q) \qquad (4.44)$$

and the right-hand side of this equivalence is itself equivalent to the proposition $x$ *being in the domain of P, all the hereditary properties of x (with respect to P) are properties of y.*

It only remains to substitute 'zero' for $x$ in (4.44) and we have a precise definition of natural number, namely the class of those classes of classes that possess all the hereditary properties of zero:

$$n \in Z_0^+ \equiv 0\ P^*n \qquad \text{Df} \qquad (4.45)$$

It is now easy to see that the membership in $Z_0^+$ of any particular number defined in the manner of (4.36), (4.37) and (4.38) is an inference from the transitivity of inclusion among classes (2.70). From (4.44) we infer that the hereditary properties of zero are included in the hereditary properties of one; those of one are included in those of two – and so on until we reach the particular number we have defined. These hereditary properties of zero are known as **inductive properties.**

The apparently converse assertion that any member of the positerity of zero can be reached by a number of steps of the kind that takes us from any member of this class to its successor is now seen to be circular, since by a number of such steps we mean precisely what takes us from zero to any member of its posterity.

We now show that succession (in the broader sense) is a serial relation (3.156), i.e., that of any two such numbers we can say that at least one belongs to the posterity of the other with respect to succession and that if each belongs to the posterity of the other then they are identical. (We prove this first.) Finally we use $P^*$ and $P$ to define the relation *strictly less than* ($<$) and show that this is strictly serial.

We begin by adopting the convenient symbolism $l + 1$ for the successor of $l$ and note in passing that this defines $l + 1$ uniquely since succession is a biunique relation. (The proof of this has been deferred to (4.86)–(4.140).

$$m = l + 1 \equiv mSl \qquad \text{Df} \qquad (4.46)$$

Next we show that if $l$, $m$ are two numbers each belonging to the posterity of the other with respect to succession, then they are identical, and conversely; in other words the relation $\leqslant$ is antisymmetric.

$$(l \leqslant m \wedge m \leqslant l \equiv (((Q \in \mathrm{her}(P) \wedge l \in Q) \Rightarrow m \in Q) \wedge ((Q \in \mathrm{her}(P) \wedge m \in Q) \Rightarrow l \in Q))) \qquad (4.47)$$

(4.44; Definition)

$$\equiv ((((x,y)(y \in Q \wedge yPx) \Rightarrow x \in Q) \wedge l \in Q) \Rightarrow m \in Q) \wedge (((x,y)(y \in Q \wedge yPx) \Rightarrow x \in Q) \wedge m \in Q) \Rightarrow (\in Q)) \qquad (4.48)$$

(4.47) and (4.41: Definition)

$$\equiv (((x,y)(y \in Q \wedge yPx) \Rightarrow x \in Q) \wedge P \in Q) \equiv ((x,y)(y \in Q \wedge yPx) \Rightarrow x \in Q) \wedge m \in Q)) \qquad (4.49)$$

(4.48) and (1.54)

$$\equiv (l \in Q \equiv m \in Q) \qquad (4.50)$$

(4.49) (3.4: $l$ for $x$, $m$ for $y$) and (3.4: $m$ for $x$, $l$ for $y$)

$$\Rightarrow (Q)(l \in Q \equiv m \in Q) \qquad (4.51)$$

(4.50) and (2.118)

$$\Rightarrow l = m \qquad (4.52)$$

(4.51), (2.116: $m$ for $Q$) and (2.118)

$$l \leqslant l \qquad (4.53)$$

(3.202 – 3.205: $P$ for $R$)

$$l = m \Rightarrow (l \leqslant m \wedge m \leqslant l) \qquad (4.54)$$

(2.162) and (1.39)

$$(l \leqslant m \wedge m \leqslant l) \equiv l = m \qquad (4.55)$$

(4.52) and (4.54)

No natural number possesses every hereditary property of its successor:

$$lPl + 1 \qquad (4.56)$$

(4.46) and (4.39)

$l \neq l + 1$ (4.57)

(4.121) and (4.140)

$l \neq l + 1 \Rightarrow ((l \leqslant l + 1)' \vee (l + 1 \leqslant l)')$ (4.58)

(4.56) and (1.15)

$l \leqslant l + 1 \Rightarrow (l + 1 \leqslant l)'$ (4.59)

(4.57) and (4.58)

$l \leqslant l + 1$ (4.60)

(4.56) and (3.206–3.216: $P$ for $R$)

$(l + 1 \leqslant l)'$ (4.61)

(4.59) and (4.60)

$l$ is the only number that possesses all the hereditary properties of $l$ but not all those of $l + 1$:

$$(l \leqslant m \wedge (l + 1 \leqslant m)') \equiv l = m$$

$P^0 \cup P|P^* \equiv P^*$ (4.62)

(Exercise 3.23: $P$ for $R$)

$(P|P^*)m \equiv l + 1P^*m$ (4.63)

(3.36) and (4.46: Df) Df

$(l = m \vee l + 1 \leqslant m) \equiv l \leqslant m$ (4.64)

(4.62), (4.63) and (2.49)

$(l \leqslant m \wedge (l + 1 \leqslant m)') \equiv l = m$ (4.65)

(4.64), (1.20) and (1.21)

The relation $\leqslant$ is transitive. This is an immediate inference from the definition and the law of successive implication:

$(l \leqslant m \wedge m \leqslant n) \Rightarrow l \leqslant n$ (4.66)

(4.44) and (1.70)

Finally, the relation $\leqslant$ is connex in the field of natural numbers:

$x = n \equiv (Q)((Q \in \text{her}(P) \wedge x \in Q)$
$\equiv (l \in Q \wedge m \in Q))$ (4.67)

Definition

$n \nleqslant l$ (4.68)

(4.67) and (4.44)

$n \leqslant m$ (4.69)

(4.67) and (4.44)

$(Q)((Q \in \text{her}(P) \wedge l + 1 \in Q) \Rightarrow n + 1 \in Q) \vee n + 1 = l$ (4.70)

(4.69) and (4.62–4.65: $n + 1$ for $m$)

$(Q)'((Q \in \text{her}(P) \wedge l + 1 \in Q) \Rightarrow n + 1 \in Q) \Rightarrow n + 1 = l$ (4.71)

(4.69) and (1.2)

$$\Rightarrow l \leqslant n \qquad (4.72)$$

(4.71) and (3.206–3.216: $P$ for $R$)

$$\Rightarrow l \leqslant m \qquad (4.73)$$

(4.72), (4.69) and (4.66)

$$(Q)'((Q \in \text{her}(P) \wedge m + 1 \in Q) \Rightarrow n + 1 \in Q) \Rightarrow n + 1 = m \qquad (4.74)$$

(4.69) and (4.62–4.65: $n + 1$ for $m$, $m$ for $l$)

$$\Rightarrow m \leqslant n \qquad (4.75)$$

(4.74) and (3.206–3.216: $P$ for $R$)

$$\Rightarrow m \leqslant l \qquad (4.76)$$

(4.75), (4.68) and (4.66)

$$((l \leqslant m)' \wedge (m \leqslant l)') \Rightarrow (Q)((Q \in \text{her}(P) \wedge l + 1 \in Q \wedge m + l \in Q) \Rightarrow n + 1 \in Q) \qquad (4.77)$$

(4.73), (4.76) and (1.15)

$$\Rightarrow (Q)((Q \in \text{her}(P) \wedge l \in Q \wedge m \in Q) \Rightarrow n + 1 \in Q) \qquad (4.78)$$

(4.77), (4.56–4.61) and (1.70)

$$(Q)'((Q \in \text{her}(P) \wedge l \in Q \wedge m \in Q) \Rightarrow n + 1 \in Q) \qquad (4.79)$$

(4.67: $x + 1$ for $x$) and (4.56–4.61)

$$l \leqslant m \vee m \leqslant l \qquad (4.80)$$

(4.78), (1.15) and (1.11)

(4.67)–(4.80) can be slightly shortened by means of the following definition:

$$l < m \equiv l + 1 \leqslant m \qquad \text{Df} \qquad (4.81)$$

From (4.72)–(4.65) an almost immediate inference is:

$$l < m \equiv (l + 1 \leqslant m \wedge l \neq m) \qquad (4.82)$$

With the help of (4.47)–(4.80) it is now fairly easy to prove the following theorems. These are left as an exercise.

*The relation $<$ is asymmetric*

$$l < m \Rightarrow (m < l)' \qquad (4.83)$$

*The relation $<$ is transitive*

$$(l < m \wedge m < n) \Rightarrow l < n \qquad (4.84)$$

*The relation $<$ is connex*

$$l \neq m \Rightarrow (l < m \vee m < l) \qquad (4.85)$$

Thus we have shown that $\leqslant$ is serial, (3.156), while $<$ is strictly serial, (3.157).

## 4.3 Peano's axioms

Until the logical foundations of mathematics were established by Frege[7] nearly a hundred years ago, it was believed that arithmetic must rest on axioms of its own. The first complete set was that due to Peano,[8] who showed that the whole of classical number-theory could be derived from the following five axioms or primitive propositions:

(i) Zero is a number.
(ii) The successor of any number is a number.
(iii) No two numbers have the same successor.
(iv) Zero is not the successor of any number.
(v) Any property belonging to zero and to the successor of any number which has that property (i.e., any hereditary property belonging to zero) belongs to all numbers.

### (a) *The primitive ideas*

It is not difficult to see the weakness of these axioms from the purely logical point of view. Zero, number and successor are undefined terms and we may easily define them in such a way as to ensure that all the primitive propositions are true, yet bear a meaning quite different to that which Peano intended. Most simply we could define zero to be what we normally designate by 'one hundred' and deny the status of number to all the traditional numbers from 0 to 99.

### (b) *Logical substitutes for the primitive ideas*

In previous paragraphs, however, we have given quite precise meanings to 'zero', 'number' and 'successor'; and we now proceed to show formally what was outlined in section 4.2(b) – that, on the basis of these definitions, the first four of Peano's axioms may be justified.

(i) Zero is a number

This follows at once from the definitions of zero section 4.2(a) and number in general (4.7).

(ii) The successor of a number is a number

If $z$ is a number and $y$ its successor, we have to show that for any two members of $y$ (say $x_1$, $x_2$) a biunique relation, $R$, exists having $x_1$ as its domain and $x_2$ as its counter-domain:

$$(x)(x \in y) \equiv (\exists w)(w \in x \wedge x \cap [w]' \in z) \quad (4.86)$$

Premiss (4.35)

$$x_1 \in y \quad (4.87)$$

Premiss

$$x_2 \in y \tag{4.88}$$

Premiss

$$(\exists w_1)(x_1 \cap [w_1]' \in z \wedge w_1 \in x_1) \tag{4.89}$$

(4.86) and (4.87)

$$(\exists w_2)(x_2 \cap [w_2]' \in z \wedge w_2 \in x_2) \tag{4.90}$$

(4.86) and (4.88)

$$(\exists Q)(z \in N(Q)) \tag{4.91}$$

Premiss (4.7)

$$((x_1 \cap [w_1]' \in z) \wedge (x_2 \cap w_2{}' \in z)) \Rightarrow ((\exists Q)(z = N(Q)) \Rightarrow (\exists Q)(x_1 \cap [w_1]' \text{ sim } Q \wedge x_2 \cap [w_2]' \text{ sim } Q)) \tag{4.92}$$

(4.4) and (1.44)

$$\Rightarrow (\exists Q)(x_1 \cap [w_1]' \text{ sim } Q \wedge x_2 \cap [w_2]' \text{ sim } Q) \tag{4.93}$$

(4.91) and (4.92)

$$\Rightarrow (\exists Q)(x_1 \cap [w_1]' \text{ sim } x_2 \cap [w_2]') \tag{4.94}$$

(4.93) and (4.11)

$$(\exists Q)(x_1 \cap [w_1]' \text{ sim } x_2 \cap [w_2]') \tag{4.95}$$

(4.89), (4.90) and (4.94)

$$x_1 \cap [w_1]' \text{ sim } x_2 \cap [w_2]' \tag{4.96}$$

(2.42) and (1.104)

$$(\exists T)(D(T) \equiv x_1 \cap [w_1]' \wedge C(T) \equiv x_2 \cap [w_2]' \wedge T \in \text{bnq}) \tag{4.97}$$

(4.96) and (4.1)

$$(i,j)(iRj \equiv iTj \vee i[w_1] \times [w_2]j) \tag{4.98}$$

Definition

$$D([w_1] \times [w_2]) \equiv [w_1] \tag{4.99}$$

(3.66) and (3.34)

$$C([w_1] \times [w_2]) \equiv [w_2] \tag{4.100}$$

(3.66) and (3.34)

$$x([w_1] \times [w_2] \mid [w_2] \times [w_1])y \equiv (x = w_1 \wedge y = w_1) \tag{4.101}$$

(3.34) and (2.188–2.212)

$$[w_1] \times [w_2] \mid [w_2] \times [w_1] \subset I \tag{4.102}$$

(4.101) and (2.162–2.166)

$$[w_2] \times [w_1] \mid [w_1] \times [w_2] \subset I \tag{4.103}$$

(4.102: $w_1$ for $w_2$, $w_2$ for $w_1$)

$$[w_1] \times [w_2] \in \text{bnq} \tag{4.104}$$

(4.102), (4.103) and (3.199)

$$(D(T) \equiv x_1 \cap [w_1]') \Rightarrow D(T) \cap D([w_1] \times [w_2]) \equiv \Lambda \tag{4.105}$$

(4.99) and (1.32)

$$(C(T) \equiv x_2 \cap [w_2]') \Rightarrow C(T) \cap C([w_1] \times [w_2]) \equiv \Lambda \tag{4.106}$$

(4.100) and (1.32)

$$(D(T) \equiv x_1 \cap [w_1]' \wedge C(T) \equiv x_2 \cap [w_2]' \wedge T \in \text{bnq}) \Rightarrow R \in \text{bnq} \quad (4.107)$$

(4.98), (4.104), (4.105) and (4.106)

$$(D(R) \equiv D(T) \cup D([w_1] \times [w_2]) \quad (4.108)$$

(4.98) and (3.66)

$$\equiv (x_1 \cap [w_1]') \cup [w_1] \quad (4.109)$$

(4.108) and (4.99)

$$\equiv x_1 \quad (4.110)$$

(4.109) and (4.89)

$$C(R) \equiv x_2 \quad (4.111)$$

Similarly

$$(\exists T)(D(T) \equiv x_1 \cap [w_1]' \wedge C(T) \equiv x_2 \cap [w_2]' \wedge T \in \text{bnq}) \Rightarrow (\exists R)(R \in \text{bnq} \wedge D(R) \equiv x_1 \wedge C(R) = x_2) \quad (4.112)$$

(4.107), (4.110) and (4.111)

$$(\exists R)(R \in \text{bnq} \wedge D(R) \equiv x_1 \wedge C(R) \equiv x_2) \quad (4.113)$$

(4.97) and (4.112)

That $y$ is the unique successor of $z$ is an almost immediate inference from the definition. We have to show that $S$ is a one–many relation (3.197).

$$y_1 S z \equiv (x)(x \in y_1 \equiv (\exists w)(w \in x \wedge x \cap [w]' \in z) \quad (4.114)$$

(4.34)

$$y_2 S z \equiv (x)(x \in y_2 \equiv (\exists w)(w \in x \wedge x \cap [w]' \in z) \quad (4.115)$$

(4.34)

$$(y_1 S z \wedge y_2 S z) \equiv (x)((x \in y_1 \equiv (\exists w)(w \in x \wedge x \cap [w]' \in z)) \wedge (x \in y_2 \equiv (\exists w)(w \in x \wedge x \cap [w]' \in z)) \quad (4.116)$$

(4.114), (4.115) and (2.13)

$$\Rightarrow (x)(x \in y_1 \equiv x \in y_2) \quad (4.117)$$

(4.116) and (1.46)

$$\Rightarrow y_1 = y_2 \quad (4.118)$$

(2.162)

$$(y_1 S z \wedge z S^{-1} y_2) \Rightarrow y_1 = y_2 \quad (4.119)$$

(4.118), (3.20) and (3.21)

$$S | S^{-1} \subset I \quad (4.120)$$

(4.119) and (3.36)

(iii) No two numbers have the same successor

We now have to show that $S$ is many–one (3.198). We recall that the field of $S$ is supposed to be confined to the class of numbers (4.41)–

(4.45); and we further suppose that the successor in question is not null:[9]

$$ySz_1 \equiv (x)(x \in y \equiv (\exists w_1)(w_1 \in x \wedge x \cap [w_1]' \in z_1) \quad (4.121)$$

(4.35)

$$ySz_2 \equiv (x)(x \in y \equiv (\exists w_2)(w_2 \in x \wedge x \cap [w_2]' \in z_2) \quad (4.122)$$

$$(ySz_1 \wedge ySz_2) \equiv (x)(x \in y \equiv (\exists w_1)((\exists w_2)(w_1 \in x \wedge w_2 \in x \wedge x \cap [w_1]' \in z_1 \wedge x \cap [w_2]' \in z_2)) \quad (4.123)$$

(4.121), (4.122) and (2.13)

$$iRj \equiv ((i \in x \cap [w_1]' \cap [w_2]' \wedge j \in x \cap [w_1]' \cap [w_2]' \wedge i = j) \vee (i\ [w_2] \times [w_1]\ j) \quad (4.124)$$

Definition

$$(w_1 \in x \wedge w_2 \in x) \Rightarrow R \in \text{bnq} \quad (4.125)$$

(4.124) and (Exercise 3.27)

$$(x \cap [w_1]' \equiv D(R)) \wedge (x \cap [w_2]' \equiv C(R)) \quad (4.126)$$

(4.124), (3.66) and (3.67)

$$(w_1 \in x \wedge w_2 \in x) \Rightarrow (x \cap [w_1]' \equiv D(R) \wedge x \cap [w_2]' \equiv C(R) \wedge R \in \text{bnq}) \quad (4.127)$$

(4.125) and (4.126)

$$\Rightarrow (\exists R)(x \cap [w_1]' \equiv D(R) \wedge x \cap [w_2]' \equiv C(R) \wedge R \in \text{bnq}) \quad (4.128)$$

(4.127) and (2.118)

$$x \cap [w_1]' \text{ sim } x \cap [w_2]' \quad (4.129)$$

(4.128) and (4.1)

$$(\ \ x)(x \in y) \quad (4.130)$$

Premiss

$$z_1 \in \text{num} \wedge z_2 \in \text{num} \quad (4.131)$$

Premiss

$$(ySz_1 \wedge ySz_2) \Rightarrow (x)(x \in y \Rightarrow (\exists w_1)(\exists w_2)(w_1 \in x \wedge w_2 \in x \wedge x \cap [w_1]' \in z_1 \wedge x \cap [w_2]' \in z_2)) \quad (4.132)$$

(4.123) and (1.143)

$$\Rightarrow ((\exists x)(x \in y) \Rightarrow (\exists x)((\exists w_1)((\exists w_2)(w_1 \in x \wedge w_2 \in x \wedge x \cap [w_1]' \in z_1 \wedge x \cap [w_2]' \in z_2)))) \quad (4.133)$$

(4.132) and (Exercise 2.7)

$$\Rightarrow (\exists x)((\exists w_1)((\exists w_2)(w_1 \in x \wedge w_2 \in x \wedge x \cap [w_1]' \in z_1 \wedge x \cap [w_2]' \in z_2))) \quad (4.134)$$

(4.131) and (4.133)

$$\Rightarrow (\exists x)((\exists w_1)((\exists w_2)(x \cap [w_1]' \in z_1 \wedge x \wedge [w_2]' \in z_2 \wedge x \cap [w_1]' \text{ sim } x \cap [w_2]'))) \quad (4.135)$$

(4.130) and (4.134)

$$\Rightarrow (\exists x)((\exists w_1)(\exists w_2)(z_1 = N(x \cap [w_1]') \wedge z_2 = N(x \cap [w_2]') \wedge N(x \cap [w_1]' = N(x \cap [w_2]')) \quad (4.136)$$

(4.134) and (4.8)

$$\Rightarrow (\exists x)((\exists w_1)((\exists w_2)(z_1 = z_2))) \quad (4.137)$$
(4.136) and (2.165)
$$\Rightarrow z_1 = z_2 \quad (4.138)$$
(4.137) and (2.42)
$$(z_1 S^{-1} y \wedge y S z_2) \Rightarrow z_1 = z_2 \quad (4.139)$$
(4.138) and (3.21)
$$S^{-1} | S \subset I \quad (4.140)$$
(4.139) and (3.71)

(iv) Zero is not the successor of any number

$$x = \Lambda \equiv (\exists w)' (w \in x) \quad (4.141)$$
(2.221)
$$\equiv (w)(w \notin x) \quad (4.142)$$
(4.141) and (2.6)
$$\Rightarrow (w)(w \notin x \vee x \cap [w]' \notin z) \quad (4.143)$$
(4.142) and (1.7)
$$x = \Lambda \not\equiv (\exists w)(w \in x \wedge x \cap [w]' \in z) \quad (4.144)$$
(4.143) and (2.6)
$$(\exists x)(x = \Lambda \not\equiv (\exists w)(w \in x \wedge x \cap [w]' \in z)) \quad (4.145)$$
(4.144) and (2.117)
$$(z)((\exists x)(x = \Lambda \not\equiv (\exists w)(w \in x \wedge x \cap [w]' \in z))) \quad (4.146)$$
(4.145) and (2.118)
$$(z)((x)'(x \in [\Lambda] \equiv (\exists w)(w \in x \wedge x \cap [w]' \in z))) \quad (4.147)$$
(4.146) and (2.5)
$$(\exists z)'((x)(x \in 0 \equiv (\exists w)(w \in x \wedge x \cap [w]' \in z))) \quad (4.148)$$
(4.147) and (4.34)
$$(\exists z)'(0Sz) \quad (4.149)$$
(4.148) and (4.35)

(v) Any hereditary property belonging to zero belongs to all numbers

The fifth of Peano's axioms is of course our definition of natural number and therefore admits of no proof.

## 4.4 Arithmetic as a consequence of Peano's axioms

It now only remains to show that the traditional arithmetic of the natural numbers can be derived from Peano's axioms. This is done in most introductions to mathematics, so that it will only be necessary to give a brief summary here. In every case proof rests upon the inductive property of the natural numbers – in other words we show that the property in question is a hereditary property with respect to succession and is a property of zero. By definition such a property belongs to all the natural numbers (4.45). This is known as **induction**.

(c) *The sum of two numbers*

We must first define the sum of two natural numbers, $l$ and $m$. It is tempting to seek a definition in terms of $L \cup M$, the union of two non-intersecting classes of $l$ and $m$. But the condition on non-intersection is clumsy, and it is easier in fact to use an inductive definition. We therefore define the successor of $l + m$ rather than $l + m$ itself. We define $l + 1$ as the successor of $l$, $l + 2$ as the successor of $l + 1$, and in general $l + (m + 1)$ as the successor of $l + m$. As 0 is not the successor of any number, we must first define $l + 0$ as $l$. Although $l$, $m$ are classes of classes, it is convenient here (as in (4.47) et seq.) to identify them not by the sign of equivalence but by that of identity or, as it is termed in this context, **equality**.

The formal definitions, including the stipulation that $l$, $m$ are numbers, are as follows:

$$(l + 0 = m) \equiv (l \in Z_0^+ \wedge l = m) \qquad \text{Df} \qquad (4.150)$$
$$(ySm \equiv z = l + y) \equiv (m \in Z_0^+ \wedge zS(l + m)) \qquad \text{Df} \qquad (4.151)$$

Putting $m = 0$, $l + 1$ is the successor of $l$ – a convention we have already adopted (4.46).

We now show that on the basis of these definitions, the ordinary laws of addition hold good.

(i) The associative law of addition

$$(l + m) + n = l + (m + n)$$

$$((l + m) + n) + 1 = (l + m) + (n + 1) \qquad (4.153)$$
(4.151: $l + m$ for $l$, $n$ for $m$)
$$(l + (m + n)) + 1 = l + ((m + n) + 1) \qquad (4.154)$$
(4.151: $m + n$ for $m$)
$$(m + n) + 1 = m + (n + 1) \qquad (4.155)$$
(4.151: $m$ for $l$, $n$ for $m$)
$$(l + (m + n)) + 1 = l + (m + (n + 1)) \qquad (4.156)$$
(4.154) and (4.155)
$$(l + m) + n = l + (m + n) \Rightarrow (l + m) + (n + 1) = l + (m + (n + 1)) \qquad (4.157)$$
(4.153) and (4.156)
$$(l + m) + 0 = l + m \qquad (4.158)$$
(4.150: $l + m$ for $l$)
$$l + (m + 0) = l + m \qquad (4.159)$$
(4.150: $m$ for $l$)
$$(l + m) + 0 = l + (m + 0) \qquad (4.160)$$
(4.158) and (4.159)

Thus $(l + m) + n = l + (m + n)$ is a hereditary property of $n$, (4.157) and a property of zero, (4.160); so it is a property of every natural number.

(ii) The commutative law of addition

$$l + m = m + l$$

Lemma $\quad 1 + m = m + 1$

$$(1 + m) + 1 = 1 + (m + 1) \qquad (4.162)$$
(4.151: 1 for $l$)

$$1 + m = m + 1 \Rightarrow 1 + (m + 1) = (m + 1) + 1 \qquad (4.163)$$
(4.162) and (2.162)

$$1 + 0 = 1 \qquad (4.164)$$
(4.150: 1 for $l$)

$$1 = 0 + 1 \qquad (4.165)$$
(4.151: 0 for $m$)

$$1 + 0 = 0 + 1 \qquad (4.166)$$
(4.164) and (4.165)

Thus $1 + m = m + 1$ is an inductive property, 'true' for all natural numbers, $m$: (4.163) and (4.166). This proves the lemma.

$$(l + m) + 1 = l + (m + 1) \qquad (4.167)$$
(4.151) and (4.46)

$$(m + l) + 1 = m + (l + 1) \qquad (4.168)$$
(4.167: $l$ for $m$, $m$ for $l$)

$$l + m = m + l \Rightarrow m + (l + 1) = l + (m + 1) \qquad (4.169)$$
(4.167) and (4.168)

$$\Rightarrow m + (l + 1) = l + (1 + m) \qquad (4.170)$$
(4.170) and lemma

$$\Rightarrow m + (l + 1) = (l + 1) + m \qquad (4.171)$$
(4.170) and (4.152)

$$(0 + m) + 1 = 0 + (m + 1) \qquad (4.172)$$
(4.151: 0 for $l$)

$$0 + m = m \Rightarrow 0 + (m + 1) = m + 1 \qquad (4.173)$$
(4.172) and (2.162)

$$0 + 0 = 0 \qquad (4.174)$$
(4.150: 0 for $l$)

thus $\quad 0 + m = m + 0$ is inductive on $m$ $\quad$ (4.175)
(4.173) and (4.174)

and $\quad l + m = m + l$ is inductive on $l$ $\quad$ (4.176)
(4.171) and (4.175)

This method of proof is known as **double induction**.

(iii) The uniqueness of the difference of two numbers

$$l + n = m + n \equiv l = m$$

$$(l + n = m + n \equiv l = m) \Rightarrow (l + (n + 1) = m + (n + 1) \equiv l = m) \quad (4.178)$$

(4.151) and (4.121)–(4.140)

$$l + 0 = m + 0 \equiv l = m \quad (4.179)$$

(4.150: $m$ for $l$)

thus $l + n = m + n \equiv l = m$ is inductive on $n$ (4.180)

(4.178) and (4.179)

(b) *The product of two numbers*

This is defined in a manner analogous to the sum, that is inductively. The general definitions are:

$$l.0 = 0 \equiv l \in Z_0^+ \qquad \text{Df} \quad (4.181)$$

$$l.y = l.m + l \equiv (m \in Z_0^+ \wedge ySm) \qquad \text{Df} \quad (4.182)$$

It is here understood that the operation of multiplication is logically prior to that of addition, i.e., in the latter definition $l.m$ stands for $(l.m)$. From this definition putting $m = 0$ we infer at once:

$$l.1 = l \quad (4.183)$$

Proofs of the traditional rules follow.

(i) The distributive law

$$l.(m + n) = l.m + l.n$$

$$l.(m + (n + 1)) = l.((m + n) + 1) \quad (4.185)$$

(4.182)

$$= l.(m + n) + l \quad (4.186)$$

(4.182)

$$l.(m + n) = l.m + l.n \Rightarrow l.(m + (n + 1)) = (l.m + l.n) + l \quad (4.187)$$

(4.186) and (2.162)

$$\Rightarrow l.(m + (n + 1)) = l.m + (l.n + l) \quad (4.188)$$

(4.187) and (4.152)

$$\Rightarrow l.(m + (n + 1)) = l.m + l.(n + 1) \quad (4.189)$$

(4.188) and (4.182)

$$l.(m + 0) = l.m \quad (4.190)$$

(4.150)

$$l.m + l.0 = l.m \quad (4.191)$$

(4.150) and (4.181)

$$l.(m + 0) = l.m + l.0 \quad (4.192)$$

(4.190) and (4.191)

thus $l.(m + n) = l.m + l.n$ is inductive on $n$ (4.193)

(4.189) and (4.192)

(ii) The associative law

$$(l.m).n = l.(m.n)$$

$$(l.m).(n + 1) = (l.m).n + l.m \tag{4.195}$$

(4.182: $l.m$ for $l$, $n$ for $m$)

$$(l.m).n = l.(m.n) \Rightarrow (l.m).(n + 1) = l.(m.n) + l.m \tag{4.196}$$

(4.195) and (2.162)

$$\Rightarrow (l.m).(n + 1) = l.(m.n + m) \tag{4.197}$$

(4.196)

$$\Rightarrow (l.m).(n + 1) = l.(m.(n + 1)) \tag{4.198}$$

(4.197) and (4.182)

$$(l.m).0 = 0 \tag{4.199}$$

(4.181)

$$l.(m.0) = 0 \tag{4.200}$$

(4.181) repeatedly

$$(l.m).0 = l.(m.0) \tag{4.201}$$

(4.199) and (4.200)

thus $(l.m).n = l.(m.n)$ is inductive on $n$ (4.202)

(4.198) and (4.201)

(iii) The commutative law

$$l.m = m.l$$

We again employ double induction, but in a slightly more elaborate form:

$$0.(m + 1) = 0.m + 0 \tag{4.204}$$

(4.182: 0 for $l$)

$$= 0.m \tag{4.205}$$

(4.204) and (4.150)

$$0.m = 0 \Rightarrow 0.(m + 1) = 0 \tag{4.206}$$

(4.205) and (2.162)

$$0.0 = 0 \tag{4.207}$$

(4.181: 0 for $l$)

Thus $0.m = 0$ is inductive on $m$ (4.208)

(4.206) and (4.207)

$$(l + 1).(m + 1) = (l + 1).m + (l + 1) \tag{4.209}$$

(4.182: $l + 1$ for $l$)

$$(l + 1).m = l.m + m \Rightarrow (l + 1).(m + 1) = (l.m + m) + (l + 1) \tag{4.210}$$

(4.209) and (2.162)

$$\Rightarrow (l + 1).(m + 1) = (l.m + l) + (m + 1) \tag{4.211}$$

(4.210) and (4.161) repeatedly

$$\Rightarrow (l+1).(m+1) = l.(m+1) + (m+1) \qquad (4.212)$$

(4.201) and (4.182)

$$(l+1).0 = 0 \qquad (4.213)$$

(4.181)

$$l.0 = 0 \qquad (4.214)$$

(4.181)

$$(l+1).0 = l.0 + 0 \qquad (4.215)$$

(4.213), (4.214) and (4.170)

Thus $(l+1).m = l.m + m$ is inductive on $m$ (4.216)

(4.212) and (4.215)

$$l.m = m.l \Rightarrow (l+1).m = m.l + m \qquad (4.217)$$

(4.216) and (2.162)

$$\Rightarrow (l+1).m = m.(l+1) \qquad (4.218)$$

(4.217) and (4.182: $l$ for $m$, $m$ for $l$)

$$0.m = m.0 \qquad (4.219)$$

(4.208) and (4.181)

Thus $l.m = m.l$ is inductive on $l$ (4.220)

(4.218) and (4.219)

(iv) The uniqueness of the quotient

$$((l.n = m.n) \wedge (n \neq 0)) \Rightarrow l = m \qquad (4.221)$$

If we replace $n$ by $p + 1$ we have a proposition that is inductive on $p$. The details are left as an exercise.

---

**Exercises**

4.1 Show that the relation $<$ is asymmetric, transitive and connex.

4.2 Show that $l.(p+1) = m.(p+1) \Rightarrow l = m$.

4.3 The extension of the basic laws of algebra to cover the operation of indices is achieved by means of the following definitions:

$$l \uparrow 0 = 1 \equiv l \in Z_0^+ \qquad \text{Df} \qquad (1)$$

$$l \uparrow y = (l \uparrow m).m \equiv (m \in Z_0^+ \wedge ySm) \qquad \text{Df} \qquad (2)$$

On the basis of these definitions show that:

(a) $l \uparrow 1 = l$
(b) $l \uparrow (m+n) = (l \uparrow m).(l \uparrow n)$
(c) $(l \uparrow m) \uparrow n = l \uparrow (m.n)$
(d) $(((l \uparrow n) = (m \uparrow n)) \wedge n \neq 0) \Rightarrow l = m$

---

**Notes on chapter 4**

1. Throughout this chapter we speak of the 'properties' of classes rather than the classes to which they belong – merely to avoid the ugly formulation 'classes of classes of classes' etc.
2. Strictly speaking, this definition excludes the number zero since alone among classes the null class is not similar to itself. However once zero is defined, this definition is supposed modified so as to include it.
3. We have an interest in so doing. It will be recalled (section **2.5**(d) that where infinite classes are concerned the formation of classes of classes presents difficulties.
4. This particular signification of $S$ and that of $P$ (the predecessor relation) is retained throughout the chapter.
5. See Note 4.
6. We omit reference to the domain of $P$ here and throughout the rest of the chapter, since $l, m, n$ are assumed to be natural numbers.
7. Gottlob Frege (1848–1925) in his *Begriffsschrift* (1879) – rediscovered by Russell more than 20 years later.
8. Guiseppe Peano, (1858–1932). They were first published in 1889.
9. This premiss is essential to the argument – it appears in the tenth line – and to the assumption made in (4.56)–(4.61) that no number is its own successor. For if the number $y$ whose successor is sought is the number of the universal class, then its successor is the null class – and so is *its* successor. But this can only happen if the number of individuals in the universe is finite.

**Definitions used in chapter 4**

(59) $P$ sim $Q$ for $(\exists R)((P \equiv D(R)) \wedge (Q \equiv C(R)) \wedge R \in \text{bnq})$
(60) $Q \in N(P)$ for $P$ sim $Q$
(61) $x \in$ num for $(\exists P)(x = N(P))$
(62) $x \in 0$ for $x = \Lambda$
(63) $ySz$ for $(x)(x \in y \equiv (\exists w)(w \in x \wedge x \cap [w]' \in z))$
(64) $x \in 1$ for $(\exists w)(w \in x \wedge x \cap [w]' \in 0)$
(65) $xPy$ for $ySx$
(66) $x \leqslant$ for $xP^*y$
(67) $Q \in$ her $(R)$ for $C(R,Q) \subset Q$
(68) $n \in Z_0^+$ for $0 \leqslant n$
(69) $m = l + 1$ for $mSl$
(70) $l < m$ for $l + 1 \leqslant m$
(71) $l + 0 = m$ for $l \in Z_0^+ \wedge l = m$
(72) $ySm \equiv z = l + y$ for $m \in Z_0^+ \wedge zS(l + m)$
(73) $l.0 = 0$ for $l \in Z_0^+$
(74) $l.y = l.m + l$ for $m \in Z_0^+ \wedge ySm$

Note that definitions (69) and (71)–(74) may alternatively be written as follows:

(69a) $l + 1$ for $\{x \mid (\exists w)(w \in x \wedge x \cap [w]' \in l\}$
(71a) $l + 0$ for $l$
(72a) $l + m$ for $\{x \mid (y)((y \in Z_0^+ \wedge xS(l + y) \equiv mSy\}$ $(m \neq 0)$
(73a) $l.0$ for 0
(74a) $l.m$ for $\{x \mid (y)((y \in Z_0^+ \wedge x = l.y + l \equiv mSy\}$ $(m \neq 0)$

# APPENDIX

## THE LOGIC OF THE SYLLOGISM

### A.1

Traditional or Aristotelean[1] logic rests upon two fundamental ideas – relations between propositions and the forms of propositions themselves. From the point of view of modern logic the first is an outcome of the calculus of propositions; the second of the calculus of sets.

### A.2 Relations between propositions

Definitions are:

(a) $p$ is said to be **indifferent** to $q$ if:

$$(p \wedge q) \vee (p' \wedge q) \vee (p \wedge q') \vee (p' \wedge q')$$

(b) $p$ is said to be the **contrary** of $q$ if:

$$(p' \wedge q) \vee (p \wedge q') \vee (p' \wedge q') \qquad \text{(i.e., } p \Rightarrow q')$$

(c) $p$ is said to be the **sub-contrary** of $q$ if:

$$(p' \wedge q) \vee (p \wedge q') \vee (p \wedge q) \qquad \text{(i.e., } q \Rightarrow p')$$

(d) $p$ is said to be the **equivalent** of $q$ if:

$$(p \wedge q) \vee (p' \wedge q') \qquad \text{(i.e., } p \equiv q)$$

(e) $p$ is said to be the **contradictory** of $q$ if:

$$(p' \wedge q) \vee (p \wedge q') \qquad \text{(i.e., } p \equiv q')$$

(f) $p$ is said to bear the relation of **superimplication** to $q$ if:

$$(p' \wedge q) \qquad \text{(i.e., } (p \Rightarrow q) \wedge (q \Rightarrow p)')$$

(g) $p$ is said to bear the relation of **subimplication** to $q$ if:

$$(p \wedge q)' \qquad \text{(i.e., } (q \Rightarrow p) \wedge (p \Rightarrow q)')$$

In this way seven out of the fifteen disjunctive normal forms on $p, q$ are accounted for. The remaining eight comprise the two simple forms of implication:

$$(p' \wedge q') \vee (p' \wedge q) \vee (p \wedge q) \qquad \text{(i.e., } p \Rightarrow q)$$

and

$$(p' \wedge q') \vee (p \wedge q') \vee (p \wedge q) \qquad \text{(i.e., } q \Rightarrow p)$$

(these are the disjunctions of (d) and (f), (d) and (g), respectively) and the six simple assertions of the truth and falsehood of $p$ or $q$ or both. The latter are just:

$$(p \wedge q) \vee (p \vee q') \qquad \text{(i.e., } p\text{)}$$
$$(p' \wedge q) \vee (p' \wedge q') \qquad \text{(i.e., } p'\text{)}$$
$$(p \wedge q) \vee (p' \wedge q) \qquad \text{(i.e., } q\text{)}$$
$$(p \wedge q') \vee (p' \wedge q') \qquad \text{(i.e., } q'\text{)}$$
$$p \wedge q$$

and

$$p' \wedge q'$$

## A.3 Forms of propositions

Every proposition[2] is supposed to be of the type $x \in x \Rightarrow x \in Y$ in one of its quantified forms, i.e.:

$$(x)(x \in X \Rightarrow x \in Y)$$
$$(x)'(x \in X \Rightarrow x \in Y)$$

or

$$(\exists x)(x \in X \Rightarrow x \in Y)$$

The form $(\exists x)'(x \in X \Rightarrow x \in Y)$ is excluded from consideration, since it is equivalent to $(x)(x \in X \Rightarrow x \notin Y)'$ and therefore asserts $X \equiv \vee \wedge Y \equiv \wedge$.

It is the prime purpose of Aristotle's system to establish the relations that exist between these three propositions and nine others of the same type that can be obtained from them by substituting $X$ for $X'$ or $Y$ for $Y'$ or both.

The complete set of twelve propositions is set out in Table A.1. A careful study of this will show that each proposition in the second subset (v–viii) has its equivalent in the third subset (ix–xii). This means that in practice there are only eight propositional types to be considered. The first four – types A and $E$ – are known as **universals**; the second four – types $O$ and $I$ – as **particulars.**

We take the particulars in their first form, $(x)'(\ldots)$, when we wish to emphasise that they are the negations of the first four; and in their alternative form, $(\exists x)(\ldots)$ when we wish to stress their verbal equivalents.

## A.4 The square of opposition

The relations that exist between these twelve propositions (or eight if we have already taken note of equivalences) are illustrated in the so-called **square of opposition** (Figure A.1).

This asserts that $A$ (the **universal affirmative**) is the contrary of $E$ (the **universal negative**); so that $E$ is the sub-contrary of $A$.

Table A.1. Propositions

| Proposition | Verbal form | Equivalent in the calculus of sets | Traditional Classification |
|---|---|---|---|
| (i) $(x)(x \in X \Rightarrow x \in Y)$ | All *X*s are *Y*s | $X \cap Y' \equiv \wedge$ | $A(E^*)$ |
| (ii) $(x)(x \in X \Rightarrow x \in Y')$ | No *X*s are *Y*s<br>No *Y*s are *X*s | $X \cap Y \equiv \wedge$ | $E(A^*)$ |
| (iii) $(x)(x \in X' \Rightarrow x \in Y)$ | All not-*X*s are *Y*s<br>All not-*Y*s are *X*s | $X' \cap Y' \equiv \wedge$ | $a^*$ |
| (iv) $(x)(x \in X' \Rightarrow x \in Y')$ | All *Y*s are *X*s | $X' \cap Y \equiv \wedge$ | $a$ |
| (v) $(x)'(x \in X \Rightarrow x \in Y)$ | Some *X*s are not-*Y*s | $X \cap Y' \not\equiv \wedge$ | $O(I^*)$ |
| (vi) $(x)'(x \in X \Rightarrow x \in Y')$ | Some *X*s are *Y*s<br>Some *Y*s are *X*s | $X \cap Y \not\equiv \wedge$ | $I(O^*)$ |
| (vii) $(x)'(x \in X' \Rightarrow x \in Y)$ | Some not-*X*s are not-*Y*s<br>Some not-*Y*s are not-*X*s | $X' \cap Y' \not\equiv \wedge$ | $o^*$ |
| (viii) $(x)'(x \in X' \Rightarrow x \in Y')$ | Some *Y*s are not-*X*s | $X' \cap Y \not\equiv \wedge$ | $o$ |
| (ix) $(\exists x)(x \in X \Rightarrow x \in Y)$ | Some *X*s are *Y*s<br>Some *Y*s are *X*s | $X \cap Y \not\equiv \wedge$ | $I(O^*)$ |
| (x) $(\exists x)(x \in X \Rightarrow x \in Y')$ | Some *X*s are not-*Y*s | $X \cap Y' \not\equiv \wedge$ | $O(I^*)$ |
| (xi) $(\exists x)(x \in X' \Rightarrow x \in Y)$ | Some *Y*s are not-*X*s | $X' \cap Y \not\equiv \wedge$ | $o$ |
| (xii) $(\exists x)(x \in X \Rightarrow x \in Y')$ | Some not-*Y*s are not-*X*s<br>Some not-*X*s are not-*Y*s | $X' \cap Y' \not\equiv \wedge$ | $o^*$ |

Similarly $A$ bears the relation of superimplication to $I$ (the **particular affirmative**); while $I$ is the sub-contrary of $O$ (the **particular negative**). $A$ is, of course, the contradictory of $O$.

In fact, once the relations of $A$ to $E$, $I$ and $O$ have been established all the others follow directly from the law of negative implication (1.15). These three relations are the direct outcome of the forms given in the first column of Table A.1 and tautologies already obtained.

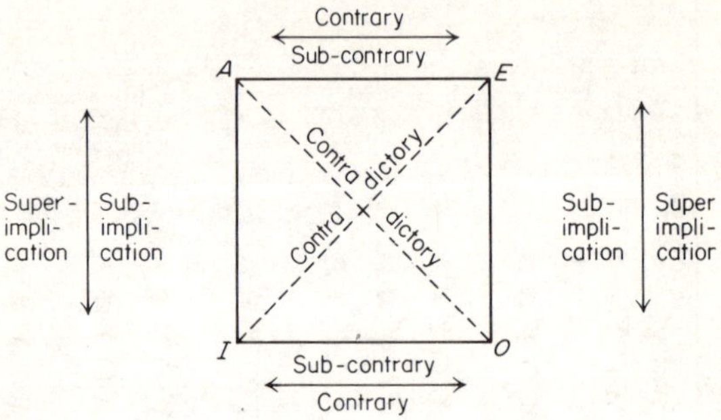


Fig. A.1

*A is the contrary of E* or $((x)(x \in X \Rightarrow x \in Y) \wedge (x)(x \in X \Rightarrow x \in Y'))'$ may be inferred from (**A.2**(b) and (1.33)).

*A bears the relation of superimplication to I* or $((\exists x)(x \in X \wedge x \in Y) \vee (\exists x)(x \in X \wedge x \in Y')) \wedge ((x)'(x \in X \Rightarrow x \in Y) \wedge (x)'(x \in X \Rightarrow x \in Y'))$ may be inferred from (A.2f) and (1.33) and (1.34), provided $X$ is not null.[3]

*A is the contradictory of O* or $(x)(x \in X \Rightarrow x \in Y) \equiv ((x)'(x \in X \Rightarrow x \in Y))'$ may be inferred from (**A.2**(e)) and (1.3).

(a) *Immediate inference*

The relation of other propositions to $A$, $E$, $O$ and $I$ is made explicit in the fourth column of table A. 1 where a change from capitals to lower case indicates an interchange of subject[4] and predicate ($X$ and $Y$), while the use of an asterisk indicates the substitution of a negative for a positive predicate ($Y'$ for $Y$). By the use of these devices, by negation and by invoking the principle of negative implication, any one of the eight different propositions may be derived from each of the others and the relation between them immediately appears.

*Worked example*

Show that $E$ is equivalent to $e$; also to $A^*$

*E* is $(x)(x \in X \Rightarrow x \in Y')$; by negative implication this is equivalent to $(x)(x \in Y \Rightarrow x \in X')$. But this is the same as *E* with *X* and *Y* interchanged, i.e., it is *e*. By inspection *E* is equivalent to *A**.

In the same way *I* is equivalent to *i*, also to *o**

(b) *Obversion*

A proposition is said to be **obverted** when it is replaced by an equivalent proposition having a complementary predicate. Thus *E** is the **obvert** of *A, no Xs are not-Ys* being the equivalent of *all Xs are Ys. A** is the obvert of *E, I** of *O* and *O** of *I.*

(c) *Conversion*

A proposition is said to be **converted** when it is replaced by another proposition equivalent to or implied by it, having as its subject the complement of the original predicate. Thus *e* is the **converse** of *E* and *i* of *I*, these being equivalents. And *i* is the converse of *A*, being implied by it.

(d) *Contraposition*

The **contrapositive** of a proposition is another proposition, equivalent or implied, having as its subject the complement of the original predicate. One way of obtaining such a proposition is by first obverting and then converting. Thus *I** being the obvert of *O* and *i** the converse of *I**, *i** is the contrapositive of *O*.

*Worked example*

Give instances of observation, conversion and contraposition

(a) *Some birds are not mammals* (*O*) obverts to
*Some birds are non-mammals* (*I**).
*Some birds are non-mammels* (*I**) converts to
*Some non-mammals are birds* (*i**).

Thus
*Some non-mammals are birds* (*i**) is the contrapositive of
*Some birds are non-mammals* (*I**) converts to

(b) *No birds are not egg-layers* (*E**) obverts to
*All birds lay eggs* (*A*).
*All birds lay eggs* (*A*) converts to
*Some egg-layers are birds* (*i*).

Thus
*Some egg-layers are birds* (*i*) is the contrapositive of
*No birds are not egg-layers* (*E**).

The latter part of this example illustrates an important point. If all birds become extinct the truth of the first proposition (*all birds lay eggs*) would be unaffected. But the inference of *i* from *A* would be invalidate because the class of birds was null. We cannot infer $x \cap Y \not\equiv \Lambda$ from $X \cap Y' \equiv \Lambda$ unless $X \not\equiv \Lambda$. If $X \equiv \Lambda$, $(x)'(x \in X \Rightarrow x \in Y)$ is a contradiction; and we can never infer a contradiction except from another contradiction. At this point the traditional logic, by failing to recognise the possibility of a null class, breaks down.

## A.5 The logic of the syllogism

Having established all possible inferences from a single proposition the traditional logic proceeds to apply the results to the syllogism. This is defined as 'the conjunction of two propositions having a common term from which a valid inference may be drawn'.

The syllogism in its simplest form we have seen (section **2.4**(a)(i) to be:

$$(X \subset Y \wedge Y \subset Z) \Rightarrow X \subset Z$$

It rests upon the fundamental theorem (2.13):

$$((x)(\phi x) \wedge (x)(\psi x)) \equiv (x)(\phi x \wedge \psi x)$$

In order to adapt the syllogism to deal with existentially quantified propositions, however, we need the additional theorem (2.37):

$$((x)(\phi x) \wedge (\exists x)(\psi x)) \Rightarrow (\exists x)(\phi x \wedge \psi x)$$

There is no inference from the conjunction of two existentially quantified propositions except Exercise 2.6:

$$((\exists x)(\phi x) \wedge (\exists x)(\psi x)) \Rightarrow (\exists x)(\phi x \vee \psi x)$$

This is of no value for syllogistic reasoning; hence the **first law of the syllogism** is *at least one of the propositions conjoined must be a universal.* The **second law** is *the proposition inferred in one premiss must be the same as that from which an inference is made in the other.* (For convenience the proposition first mentioned is generally placed first in the conjunction.)

Thus, ignoring quantifiers for the moment,

$$\left.\begin{array}{l} x \in X \Rightarrow x \in Y \\ x \in X' \Rightarrow x \in Y \end{array}\right\} \quad \text{may be conjoined with} \quad \left\{\begin{array}{l} x \in Y \Rightarrow x \in Z \\ x \in Y \Rightarrow x \in Z' \end{array}\right.$$

and

$$\left.\begin{array}{l} x \in X \Rightarrow x \in Y' \\ x \in X' \Rightarrow x \in Y' \end{array}\right\} \quad \text{may be conjoined with} \quad \left\{\begin{array}{l} x \in Y' \Rightarrow x \in Z \\ x \in Y' \Rightarrow x \in Z' \end{array}\right.$$

or, taking the first and last four forms from Table A.1:

| | | |
|---|---|---|
| *A*, *a**, *O**, *o* in the first | may be conjoined with | *A*, *A**, *O**, *O* in the second |

provided that two *o*s are not conjoined; and

| | | |
|---|---|---|
| *A**, *a*, *O*, *o** in the first | may be conjoined with | *a**, *a*, *o*, *O** in the second |

again provided that two *o*s are not conjoined.

*Worked example*

What is the conclusion from *A* in the first proposition and *E* (i.e., *A**) in the second?

The premisses are:

$$(x)(x \in X \Rightarrow x \in Y)$$
$$(x)(x \in Y \Rightarrow x \in Z')$$

The conclusion is:

$$(x)(x \in X \Rightarrow x \in Z')$$

In words:

From *All Xs are Ys and no Ys are Zs* we infer *No Xs are Zs.*

*Worked example*

What is the conclusion from *O* in the first proposition and *a** in the second?

The premisses are:

$$(\exists x)(x \in X \Rightarrow x \in Y')$$
$$(x)(x \in Y' \Rightarrow x \in Z)$$

The conclusion is:

$$(\exists x)(x \in X \Rightarrow x \in Z)$$

In words:

From *some xs are not-Ys and all not-Ys are Zs* we infer *some Xs are Zs.*

## A.6 Extensions to the logic of the syllogism

The logic of the syllogism can be extended in various ways as has already been noted. It can be extended in a fairly obvious manner to the sorites (section **2.4**(a)(iii)); it can also be extended to the **hypothetical** and **disjunctive syllogisms.** All that is required is to rewrite the arguments of chapter 1 in terms of suitably quantified propositions.

(a) *Hypothetical syllogism*

As an example of the hypothetical syllogism we take:

$$(((p \Rightarrow q) \Rightarrow (p \Rightarrow r)) \wedge (p \Rightarrow r')) \Rightarrow (p \Rightarrow q)'$$

In terms of the calculus of sets this is:

$$((X \subset Y \Rightarrow Y \subset Z) \wedge (X \subset Z)') \Rightarrow (X \subset Y)'$$

or, if we choose an existential quantifier to govern the second premiss,

$$((X \subset Y) \Rightarrow Y \subset Z) \wedge (\exists x)(x \in X \wedge x \in Z')) \Rightarrow (X \subset Y)'$$

(b) *Disjunctive syllogism*

As an example of the disjunctive syllogism:

$$(((p \Rightarrow q) \vee (p \Rightarrow r)) \wedge (p \Rightarrow r)') \Rightarrow (p \Rightarrow q)$$

$$\text{i.e. } ((X \subset Y \vee X \subset Z) \wedge (X \subset Y)') \Rightarrow (X \subset Z)$$

or, again making use of the existential quantifier,

$$((x)(x \in Y \cup Z) \wedge (\exists x)(x \in Y')) \Rightarrow (\exists x)(x \in Z)$$

It is clear that the possibilities are endless. Little is to be gained either by elaborating them or by attempting an exhaustive classification. What is important is to realise that the method of truth-tables and the rules of quantification developed in chapter 2 provide us with a means of establishing the validity or otherwise of any argument of this kind.

---

**Exercises**

A.1 Express the following premisses in the symbolism of the calculus of sets and infer as many conclusions as possible. Take as concourse *ducks in this village.*

(a) All ducks in this village that are branded $B$ belong to Mrs Bond.
(b) Ducks in this village never wear lace collars unless they are branded $B$.
(c) Mrs Bond has no grey ducks in this village. (Carroll)

A.2 Repeat question 1 with the following:

(a) Promise-breakers are untrustworthy.
(b) Wine-drinkers are very communicative.
(c) A man who keeps his promises is honest.
(d) No teetotalers are pawnbrokers.
(e) One can always trust a very communicative person. (Carroll)

A.3 Taking *birds* as concourse and $\phi x$ as *x is white*, $\psi x$ as *x is a swan*, $\chi x$ as *x is a cygnet*, express symbolically:
(a) All cygnets are swans.
(b) Not all swans are cygnets.
(c) Some cygnets are not white.
(d) All swans that are not cygnets are white.
(e) Some birds that are not white are not cygnets.

A.4 Taking A.3 (a)–(e) as premisses show how to infer:
(a) Not all swans are white.
(b) Some birds are not swans.
(c) Some swans are white.

A.5 Taking *gold glitters* as *A*, the universal affirmative, of what type is each of the following and what relation obtains between each pair?
(a) Gold glitters.
(b) Only gold glitters.
(c) All that glitters is not gold.
(d) Nothing both glitters and is gold. (London)

A.6 Repeat A.5 with the following:
(a) Only where fools rush in do angels fear to tread.
(b) Some places where fools rush in, angels fear to tread.
(c) Not everywhere that angels fear to tread do fools rush in.
(d) It is not only where angels fear to tread that fools rush in.

A.7 Render the following premisses into symbols and show how to draw the conclusion:
(a) All Cretans are either fools or knaves.
(b) Zeno is a Cretan and no fool.

---

**Notes on appendix 1**

1. Aristotle of Stagira (384–322 B.C.) was the virtual founder of traditional logic and its complete exponent. His works devoted to this subject are the *Prior Analytics*, the *Posterior Analytics* and the *Topics.*
2. The singular proposition, $x \in y$, is at first sight an exception; in fact it is merely the special case which arises when $X \equiv V$. In this case the quantifier is dropped.
3. See the last paragraph of **A.4**(d). To establish the second conclusion we do not need to show that *I implies A* is 'false' for all *X, Y*; only that *I implies A for all X, Y* is false; i.e., that for some, *X, Y, A* is 'false' and *I* is 'true'. This we can do by taking $\vee$ for *X*.

4. In the traditional logic the word ‘subject’ is used in a sense rather different from that suggested in section 2.2(a)

## SUGGESTIONS FOR FURTHER READING

R. L. Goodstein, *Boolean Algebra* (Pergamon Press): substantially extends the algebra developed in chapter 2.

R. L. Goodstein, *Mathematical Logic* (Leicester U.P.): extends the range of the whole book, especially the question of decidability.

C. L. Hamblin, *Elementary Formal Logic* (University Paperbacks): a programme of exercises designed to secure the reader’s grasp of material presented in chapters 1 and 2.

S. K. Langer, *An Introduction to Symbolic Logic* (Dover): a very thorough introduction for the general reader.

E. J. Lemmon, *Beginning Logic* (Nelson): covers most of the ground of chapters 1–3, but from a rather different standpoint.

W. van O. Quine, *Mathematical Logic* (Harvard U.P.): for the advanced student.

B. Russell, *Introduction to Mathematical Philosophy* (Allen and Unwin): still probably the best introduction to the whole subject. Written while the author was serving a prison sentence for his political activities.

L. S. Stebbing, *A Modern Elementary Logic* (University Paperbacks): deals with the subject matter of chapters 1 and 2 with the minimum of symbolism.

R. R. Stoll, *Sets, Logic and Axiomatic Theories* (W. H. Freeman): an especially good introduction for those interested in the axiomatic approach. Extends the subject matter of chapters 1–3.

R. R. Tarski, *Introduction to Logic* (O.U.P., New York): combines a lucid exposition and a plentiful supply of exercises with a penetrating analysis of the scientific method.

# INDEX